How the Martians Discovered Algebra

Explorations in Induction and the Philosophy of Mathematics

by Roger E. Bissell

Copyright and Description

*Text copyright © 2014, 2017 by Roger E. Bissell
and Bissell Words & Music, Antioch, Tennessee
All Rights Reserved*
**ISBN-13: 978-1548260620
ISBN-10: 1548260622**

This exploration of induction and philosophy of mathematics is presented as a look "under the hood" at the process of mathematical theorizing, a detailed view of how the process of induction actually works. It also provides an alternative to the mind-numbing constructs of modern logic, mathematics, and set theory, explaining the true nature of zero and empty sets and revealing the flaw in Cantor's writings on infinity.

There are original ideas here, including the author's view that zero and the empty set function as "operation blockers," as well as his explanation of why the value of the zero power of any number is always 1. The author also offers his own discovery of a new method for generating Pythagorean triples. He lays out both the deductive validation of his method and the details of his exploration of the Pythagorean equation that uncovered the relationships underlying his method.

Academics, college students, and intelligent laypersons interested in philosophy and mathematics will all find this a challenging and stimulating read. They will be rewarded with new perspectives, not only on the theoretical landscape of mathematics and logic, but also on the value in learning the mental processes involved in induction, as well as the endless opportunities for fascination and delight to be obtained from mathematical discovery.

*www.rogerbissell.com
rebissell@aol.com*

Note to readers: This book is also available in Kindle format as an e-book from Amazon.com.

About the Author, Roger E. Bissell

Roger Bissell, a musician and writer, was born in Iowa in 1948 and is happily married and a proud father and grandfather. While a youngster, he studied music fundamentals, jazz, and composition from the legendary trombonist Rex Peer, and he received a Bachelor of Science in music theory and composition, with a minor in mathematics, from Iowa State University in 1970 and a Master of Arts in music performance and literature from the University of Iowa in 1971.

Since college, Roger has been a professional musician in Nashville, Tennessee (14 years), in California (25 years), and once again in Nashville since 2010. His thousands of concerts, club gigs, live television and radio programs, and recording sessions have included appearances with Lucille Ball, Bob Hope, Henry Mancini, Jack Sheldon, The Four Freshmen, Jerry Vale, Bobby Vinton, Wayne Newton, Jay Leno, Gordon McRae, Johnny Cash, Ray Charles, B. J. Thomas, Marty Robbins, Ronnie Millsap, Chet Atkins, Floyd Cramer, Perry Como, Ray Stevens, Mannheim Steamroller, Maureen McGovern, Dolly Parton, Toni Tennille, Brenda Lee, Crystal Gayle, and the Mandrell Sisters, among others. His arranging clients have included Disneyland, The California Wind Orchestra, Glenn Campbell, Boots Randolph, and the Oak Ridge Boys.

During and after Roger's 25 years with the Disneyland Band, he has also appeared since 2002 as a regular member of the Side Street Strutters Jazz Band on community concerts around the country, as well as on pops concerts with the Phoenix, Rochester, Atlanta, Orlando, and Houston Symphony Orchestras (among others). His 1992 recording "The Art of the Duo" with pianist Ben Di Tosti was released in 2003 to much critical acclaim (including from the renowned pianist-composers Dave Brubeck and Clare Fischer). Along with his 2010 solo recording, "Reflective Trombone," it is available from CDBaby and Itunes on Muse Seeker Records. Other recordings in which Roger figures prominently are "The Secret's Out" (Earwitness, 1978), "Eli's Coming" (Nashville Jazz Machine, 1981), "This Swingin' Life" (Don Miller Orchestra, 1998), "Back to Bourbon Street" (Side Street Strutters Jazz Band, 2006), and "Shiny Stockings" (Side Street Strutters Jazz Band with Meloney Collins, 2012).

(http://www.cdbaby.com/cd/rbissell,
http://www.cdbaby.com/cd/rogerbissell)
(http://sidestreetstrutters.com/?page_id=12)

Roger has also taught and lectured on improvisation and emotional expression in music, as well as the nature of art and music more generally. Several of his published essays are posted at http://www.rogerbissell.com. Roger is also engaged in a long-term project of analyzing the emotional content of American popular songs and Classical themes, and helping listeners, performers, composers, and arrangers to apply the results of his research in order to achieve a more satisfying musical experience. He may be reached by email at rebissell@aol.com.

In addition, Roger has written extensively on philosophy and psychology, his essays appearing in *Individualist*, *Reason*, *Reason Papers*, *The Libertarian Forum*, *Objectivity*, *Journal of Consciousness Studies*, *Bulletin of the Association for Psychological Type*, *Vera Lex*, *ART Ideas*, and *The Journal of Ayn Rand Studies*. His essay "A Calm Look at Abortion Arguments" was reprinted in *Free Minds and Free Markets* (Reason Foundation, 1994), his mock transcription of a lecture by fictional composer Richard Halley was published in Edward W. Younkins's 2007 compilation, *Ayn Rand's "Atlas Shrugged": A Philosophical and Literary Companion*. Roger has also done extensive work transcribing the lectures of various philosophers. In particular, his supervised transcription of Nathaniel Branden's recorded lectures was the basis for their publication in 2009 as *The Vision of Ayn Rand: The Basic Principles of Objectivism*, and his transcription of Barbara Branden's recorded lectures were published in 2017 as *Think as if Your Life Depends on It: Principles of Efficient Thinking and Other Lectures*.

Dedication, Inspirational Quotations

To the two greatest inducers of all time:
Aristotle (384-322 B.C.E.)
&
Srivinasa Ramanujan (1887-1920)

* * *

All science is of the universal and necessary.
(Aristotle, *Posterior Analytics*)

Now, from the very fact that Aristotle and the Scholastics considered it possible to reach a truth about *"all,"* actual and possible, known and unknown, by an acquaintance with *"some,"* they must have recognized a method of ascent to the *"all,"* other than enumeration. And so they did: viz., the method nowadays known as *Physical or Scientific Induction*.
(Peter Coffey, 1912, *The Science of Logic, An Inquiry into the Principles of Accurate Thought and Scientific Method*)

…an aspect of mathematics which is as important as it is rarely mentioned: mathematics appears here as a close relative to the natural sciences, as a sort of "observational science" in which observation and analogy may lead to discoveries.
(George Polya, 1965, *Mathematical Discovery*)

[T]he role of inductive evidence in mathematical investigation is similar to its role in physical research…[M]athematics is, in several respects, the most appropriate experimental material for the study of inductive reasoning…[I]nductive research may be useful in mathematics in another respect that we have not yet mentioned…From the intent examination of a particular case a general insight may emerge.
(George Polya, 1954, *Mathematics and Plausible Reasoning*)

It [the theory of numbers] is also a popular topic among amateur mathematicians (who have made many contributions to the field) because of its accessibility: it does not require knowledge of higher mathematics...No special training is needed – just high school mathematics, a fondness for figures and an inquisitive mind...The path is endless, but many rewards are offered along the way. One could do worse than follow the gleam of numbers.
(C. Stanley Ogilvy and John T. Anderson, 1966, *Excursions in Number Theory*)

Even in the published notebooks, you can see Ramanujan giving concrete numerical form to what others might have left abstract – plugging in numbers, getting a feel for how functions "behaved." Some pages...look less like mathematical treatise, more like the homework assignment of a fourth-grader. Numerical elbow grease it was. And he put in plenty of it. One Ramanujan scholar, B. M. Wilson, later told how Ramanujan's research into number theory was often "preceded by a table of numerical results, carried usually to a length from which most of us would shrink"...Ramanujan was doing what great artists always do – diving into his material. He was building intimacy with numbers, for the same reason that the painter lingers over the mixing of his paints, or the musician restlessly practices his scales. And his insight profited. He was like the biological researcher who sees things others miss because he's there in the lab every night to see them...His successes did not come entirely from flashes of inspiration. It was hard work. It was full of false starts. It took time.
(Robert Kanigel, 1991, *The Man Who Knew Infinity*)

Contents

Preface

Introduction: Confessions of a Would-Be Mathematician 1

    Enthusiastic recommendations 13

Chapter 1: "Speed Math" 15

    How I discovered "speed math" 15

    Further references 19

    Postscript – sums of counting numbers 19

    Postscript 2 - what is the point of speed math? 22

Chapter 2: A New Procedure for Deriving Pythagorean Triples 23

    Part I: Derivation of a new procedure for generating Pythagorean triples 23

    Part II: Comparison of the author's method with the traditional procedure for generating Pythagorean Triples 26

    For more information on formulas for generating Pythagorean triples 27

Chapter 3: What's the Deal with X-to-the-Zero Power? A Brief Note on the Non-Fiction of Zero Exponents 29

    Postscript – comparing this perspective to the standard approach 33

    Postscript 2 – extending this approach to fractional exponents 35

    Postscript 3 – further clarifications 37

Chapter 4: How the Martians Discovered Algebra 41

    Postscript – light speed and mass-energy equivalence 43

    Postscript 2 – is the speed of light an absolute limit? 47

Chapter 5: Mathematics as an Inductive Science 51

    Postscript – Peikoff, Groarke, and Kornblith on induction 56

    Postscript 2 – Joseph and Coffey on induction 60

Chapter 6: Equations as Propositions – Using Aristotle to Prove Euclid 65

Equations as propositions 67

Categoricals vs. conditionals in logical arguments 72

Postscript – when not to use conditionals 81

Chapter 7: Much Ado about Nothing – Zero as an "Operation Stopper" 85

Postscript – is zero-to-the-zeroth power undefined? 105

Postscript 2 – the (non-)effect of the author's view of zero on long multiplication, etc 107

Postscript 3 – mercury, money, and metaphysics 108

Chapter 8: More ado about nothing – The Inconvenient Fiction of "Empty Sets" 113

Postscript – empty sets as operation-blockers 126

Postscript 2 – are numbers merely imaginary? 128

Postscript 3 – other examples of non-existence as an operation blocker 130

Postscript 4 – is the "operation blocker" idea just a semantic quibble? 132

Chapter 9: …and Everything! Thoughts on Induction and Infinity 133

Postscript – the intellectual and moral bankruptcy of modern math and science 142

Postscript 2 – arguments about infinity 143

Appendix: Studies on the Pythagorean Theorem 145

Part I: Two infinite classes of right triangles with integral-length sides 145

Part II: Generalizing to an infinite number of infinite classes of right triangles with integral-length sides 153

Part III: Extending the generalization even further 159

Part IV: Limits on the ratios of the sides of a right triangle 167

Conclusion 173

Postscript – comparison with other formulas for generating Pythagorean triples 173

Preface

For a number of years, I have been writing, gathering, and editing essays for books on a variety of subjects, including guidance for musicians and logic students, as well as thoughts on political philosophy, logic, epistemology, aesthetics, music, and various specific issues in philosophy. Having so many "irons in the fire," I felt gridlock begin to set in, so I asked myself: "What do I really, really most want to publish first?" The answer was not even among my (supposed) top six priorities: the book of essays on mathematics and the process of induction which you are now reading.

Happily, putting this book together, first as a Kindle e-book and then as a CreateSpace paperback, was not so much a matter of planning and writing, as of "harvesting" – in essence, scouring my computer files for appropriate material. After that, it was a simple matter to edit them into readable form and then take advantage of the no-cost self-publishing opportunities provided by Amazon.com. (So, thank you, Amazon!)

The various chapters appear in chronological order. The "easier," more "reader-friendly" essays are Chapters 1-5 and the quasi-autobiographical Introduction. I suspect that "math geeks" will be more at home in this part of the book. Chapters 6-9, on the other hand, are more abstract in nature and will probably appeal more to "philosophy freaks." However, "your mileage may vary," as they say. (And if the reader is a geek *and* a freak, like the author, then "it's all good.")

I'm particularly excited about Chapter 2 and the Appendix, which present my original discovery regarding generation of Pythagorean triples, and Chapter 6 which includes a foretaste of what is to come in the book on logic and epistemology on which I am currently working. (The Appendix, which is actually a longer version of Chapter 2, is hardly reader-friendly, unless you're in love with page after page of detailed calculations! It's included mainly to "show my work" and to provide interested readers with an in-depth look at an extended inductive process.)

On a more personal note, I deeply appreciate the feedback and encouragement I've received from my friends Dennis Edwall, Allen Tupker, Milo Schield and Debra Martin, and my wife, Elizabeth Bissell, as well as from the many individuals taking part in the aforementioned online discussions. I couldn't have done it without you, folks!

Also, I'm truly grateful to have been able to explore some of these ideas with other independent scholars and writers and then share my thoughts on them with a far-flung readership by means of electronic and print self-publishing, and with very little technical difficulty or financial risk. Chapters 3, 6, 7, 8, and 9, in particular, could not have happened without the Internet. They grew out of my involvement with online discussion groups, including Objectivist Living and Atlantis, as well as cyber-seminars directed by David Kelley and Chris Sciabarra.

Finally, I want to state for the record that, while I am a long-time fan of Ayn Rand and her philosophy of Objectivism, as may be gathered from comments I make from time to time in this book, I am not endorsed by or affiliated with any official Objectivist organization. Nor am I recognized or accepted as an "Objectivist" philosopher. Instead, I am simply working independently within the Objectivist philosophical framework – on my own projects, on my own time.

Thus, these essays are not intended in any way to be a rigorous, systematic, and integrated presentation of Objectivist ideas on mathematics and induction, nor even more generally an attempt to promote Rand's ideas, but simply some thought-provoking and entertaining reading for those interested in these subjects. They are just the first of a series of many such windows into the world of my intellectual interests and insights, which sometimes stem from and often intersect with Rand's philosophy, but for which and in which I take complete responsibility and pride.

Comments, questions, and corrections are always welcome at rebissell@aol.com.

Happy reading!
Roger Bissell, Antioch, Tennessee
May 2014
Update #1: July 2017

Introduction: Confessions of a Would-Be Mathematician

For a couple of years in the mid-1960s, I seriously considered becoming a mathematician. By the fall of 1968, however, I had abandoned my double major (mathematics and music), casting my academic and career lot solely with music. Even now, semi-retired, I'm still a professional musician, and I have been, non-stop, ever since my first paid music job in June of 1963.

Music didn't actually become my *livelihood* until I finished college in May of 1971 and moved to Nashville, Tennessee. But even if you don't count the eight years I made music in order to support my college education, I have been a professional musician for nearly 46 years now – certainly long enough to have had second thoughts about this career choice.

Confession #1: So, what might I have been, had I chosen differently? Most likely, either a psychologist, or a philosopher – or a mathematician.

A psychologist, because I have always tried to understand and deal with people's hang-ups, my own included. A philosopher, because I have always wondered why the world is in the mess it is, and how to fix it. The common denominator of these two deep interests is not only understanding the causes of important problems, but also solving them.

But why a mathematician? Not really for any moral cause, nor even for any obviously practical reason. Oh, sure, I balance my own checkbook and keep the family budget records. I do my own and others' taxes. But I have no special love for any of these uses of math.

What about the pleasure of solving problems? Well, I do enjoy challenges. And I do get a kick out of using my ingenuity and insight, almost as combat weapons, to defeat "the enemy," the *problem*. But this is not the heart of it.

What about the supposed deep connection between math and music? After all, I was a double major my second year in college. Was I drawn by the so-called Pythagorean "bridge" between the two fields? No way. I am well aware of the "music of the spheres" and the mathematical ratios in tones and harmonies. But these have nothing to do with why I love either subject.

I love music because it lets me exercise my ingenuity and senses of humor and beauty in creating patterns. I love math for similar reasons: using ingenuity to *build* patterns.

The way I like to do music is artistic and inventive, creating expressive patterns of musical notes that didn't exist before. Although math (the way I like to do it) similarly involves creating patterns of numbers that didn't exist before, it is less like invention and more like detective work, finding relationships that already exist by creating patterns that reveal those relationships.

So, if I had been a mathematician, it would have been for the thrill of the chase, the joy of discovering the elusive pattern, the pleasure of capturing the wily theorem. Like a Sherlock Holmes of numbers.

Mathematics is also very similar to science. In the last chapter of *Mathematical Discovery* (1965), Stanford professor George Polya wrote about "Guessing and Scientific Method." He spoke of:

> an aspect of mathematics which is as important as it is rarely mentioned: mathematics appears here as a close relative to the natural sciences, as a sort of "observational science" in which observation and analogy may lead to discoveries.

That's actually what I like *best* about mathematics: the opportunity to be a *scientist*, to explore and discover what exists.

In general, as a matter of fact, I am powerfully drawn to investigating, to finding the hidden, deep truth – whether on the concrete level (tracing "lost" people in the past or present, understanding what makes a person "tick") or the abstract level (philosophy, psychology, mathematics, etc.).

It's not just curiosity, wanting knowledge and understanding. It's more like being a *hunter-gatherer*, stalking the elusive, hidden fact or essence. My various strengths, such as logic or ingenuity or abstract thinking ability, all seem channeled into my desire to track down something and say, "Gotcha!"

But when did this love first manifest itself? And why didn't I follow through with it as a career? And how have I dealt with this unrequited love since then?

As far back as I can remember, I've always been good at math. I got A's all through grade school and high school, and I scored at or near the 99th percentile in the math sections of all the achievement tests I took.

But I haven't always had a clear sense that there was something important and special about math. That didn't happen until the 10th grade, when I took geometry. For the first time, it seemed, my mind had come alive. (Or at least, part of my mind; my writing skills didn't really ignite until my freshman and sophomore years of college.)

What was so important about geometry? For me, it was learning how to think. Geometry is logical. Not that algebra isn't. But in geometry, you learn how to reach conclusions that may require many steps of proof.

Why was geometry so special? Because it's visual. You're not just manipulating numbers or letters. Instead, you deal with lines, shapes, angles, and figures. Being a doodler, I felt very much at home in geometry.

Trigonometry, which I took the fall of my senior year, was part algebra, part geometry. The main use of proof was in deriving formulas. The same was true for analytical geometry and calculus, the last half of my senior year.

But something really amazing happened during my senior year. (Amazing, in retrospect. At the time, it just seemed like fun.) And it was not part of the high school math curriculum either. I discovered *number theory*.

Of course, I didn't know it was number theory. I didn't realize there was a whole branch of math based on exploring the relationships between numbers. I just stumbled onto it one day.

For some time, I had been intrigued by the fact that a given number's double ($2n$) and square (n^2) are almost always different. The only except is when the number is 2, in which case its square and double are both 4. Why is that? I wondered.

I decided to make two columns listing the doubles and squares side by side. The first thing I noticed was that for numbers greater than 2, the square was larger. Moreover, the difference between them increased rapidly, as the numbers became larger.

Was there some relationship between them? I wondered. And then, a pattern seemed to leap out at me.

Just looking at the columns of numbers, it suddenly struck me that, for any number, its square (in the second column) was one more than the sum of the next smaller number's square and double (in the first and second columns). Or, to put it another way, if you add 1 to the sum of the square and double of any number, you get the square of the next larger number.

An application immediately suggested itself. Suppose you want to square a number. Simply add the square and the double of the next smaller number, plus one. Speed math! (If you know the smaller square, of course.)

Once my initial excitement died down a bit, I wondered if there weren't some general, symbolic expression for this discovery. Of course, there was. Every freshman algebra student has seen it: $(x + 1)^2 = x^2 + 2x + 1$. That realization took some of the wind out of my sails. What I had discovered wasn't so special after all.

What I didn't realize then was that the "what" of my discovery wasn't nearly so important as the "how," the method by which I discovered the "what." It wasn't until a couple of years later, when reading Ayn Rand's *Introduction to Objectivist Epistemology*, that I realized the importance of this discovery.

Rand explained that there are two directions in which our minds work in grasping knowledge. Upward from particulars, or induction. And downward from more general knowledge, or deduction.

Introduction: Confessions of a Would-Be Mathematician

I realized then that these two different pathways also apply to discovery of principles and equations in mathematics, such as my initial inductive discovery in number theory which "reinvented" the deductive "wheel" from freshman algebra.

In the years since, I have inductively discovered other "speed math" techniques. I discuss some of them in Chapter 1. Yes, I realize I could have deduced them using algebra. But would I have known what I was looking for?

For instance, I discovered that the square of any number minus one is equal to the product of the next larger and next smaller number – and more generally, that the square of any number minus the square of any other number is equal to the product of the sum and the difference of those two numbers. Here are two examples:

$$99^2 - 1^2 = 9801 - 1 = 9800 = 100(98)$$
$$97^2 - 3^2 = 9409 - 9 = 9400 = 100(94).$$

I have used these (and similar) techniques numerous times, and I can vouch for their power and practical value. Yet, I doubt that they would have occurred to me merely by knowing and reflecting on the facts that $x^2 - 1 = (x + 1)(x - 1)$ and $x^2 - a^2 = (x + a)(x - a)$.

In general, induction just seems to work better for discovery, for forming hypotheses. (And yes, that generalization is another inductive conclusion!)

Deduction, on the other hand, seems more effective at *proving* those conclusions. I have seen this principle at work in physics, too. See the Postscript to Chapter 4, my short story about how the Martians discovered algebra.

So – **Confession #2** – while I love deductive thinking and proof, and I find them very necessary in doing math, my real love affair has been with *intuitive induction*. The raw, unrefined search for patterns is what really turns me on.

But as I said, I was unaware that number theory was an option, something I might specialize in. I just knew that, more than ever, math looked like a fun thing to do. This definitely reinforced my decision to major in math (along with the paucity of scholarship offers from music schools).

My freshman year at Iowa State University was pretty uneventful. Despite a busy schedule of music activities, I chugged through three quarters of calculus with A's and B's, and I aced Beginning Symbolic Logic.

I started my sophomore year with a double major – and great expectations. Then all hell broke loose. I met my Waterloo: Abstract Algebra and Intermediate Symbolic Logic.

Part of my trouble was the distraction from music and outside interests. But the main problem was what I call the "reality factor." My math studies had finally taken me past all the familiar signposts. I was no longer in the real world!

In Abstract Algebra, we no longer had nice friendly problems like: if train A is traveling west at 50 miles an hour and train B, 200 miles away, is traveling east at 75 miles an hour, how long before they collide? (For details, see *Atlas Shrugged*!)

And in Intermediate Symbolic Logic, we were no longer concerned with proving a conclusion, given certain assumptions, or with analyzing an argument to see if it contained a fallacy.

Instead, to my bewilderment and frustration, I found myself flailing about in alternate systems of arithmetic and logic. Frameworks where 2 times 3 is not necessarily equal to 3 times 2, and where A is not necessarily A. Or so it seemed.

With nothing solid to hold on to, I floundered about like a non-swimmer in the deep end. I prayed for the end of the quarter. It finally came, with C's in both courses, and my self-esteem and G.P.A. both took a nosedive.

This personal and scholarly disaster was in stark contrast to my glowing successes in music performance, theory and composition. Even two B's in Advanced Calculus were not enough to dispel the gloom that had fallen over my seemingly doomed future in math.

Bottom line: in the spring of 1968, I dropped the math part of my double major like a hot potato and finished my B.S. degree with honors in music.

Introduction: Confessions of a Would-Be Mathematician

(I completed my math minor by eking out a B in something I vaguely remember as "Discrete, Finite Mathematics.")

Unaware that the mathematical "potato" had sprouts that would someday grow back to the surface, I turned instead to other areas. Philosophy and psychology became my principal intellectual outlets.

But time after time, along the way, my pattern-lust continued to assert itself. Not just in playing jazz and in writing songs and musical arrangements, but in various other ways, such as:

- ✓ summarizing eight years of income and work frequency figures of my recording career, relating it to the ups and downs of the economy and the music business;
- ✓ comparing 10 years of trends in employment and wages in the recording business and the general economy;
- ✓ analyzing 12 years of taxing, spending, and employment figures for the Metro Nashville government and public schools; and
- ✓ perusing Nashville election and referendum returns for 1976 through 1982.

I used the data from these explorations in various practical ways, including:

- ✓ a 1979 master class to would-be professional trombonists,
- ✓ an Open Letter to my musical colleagues in 1984 ("Union Scale is Killing Our Work") arguing that recording scale should be drastically lowered,
- ✓ repeated recommendations during the early 1980s that taxes and spending in Metro Nashville should be drastically lowered, and
- ✓ an assessment of the effectiveness of targeted campaigning in Metro Nashville.

(I also write about these projects in my forthcoming e-books *What They Didn't Teach Me in Music School* and *The Experience of Liberty*.)

In each case, I was trying to help solve problems for myself and others in my profession and my community. But more importantly to my ego and my psyche, I was hot on the trail of patterns hidden in the numbers.

My love of mathematics continued to assert itself in these underground sorts of ways. Finally in 1988, an article on a supposed proof of Fermat's Last Theorem hooked me again, this time for good.

Back in 1637, Pierre de Fermat claimed to have proved that the only integral solution to the equation $x^n + y^n = z^n$ was $n = 2$. Since he did not record his proof, mathematicians have struggled in vain ever since to prove his claim.

I first heard of this theory about 51 years ago back in freshman algebra, in the spring of 1963. We were studying its Pythagorean form, the equation $x^2 + y^2 = z^2$. Our teacher challenged us (facetiously, of course) to find an integral solution where $n > 2$.

When I returned to the problem 25 years later in 1988, I took a different approach. Rather than trying to prove *that it didn't* work for $n > 2$, I tried to find out what it is about the equation *that lets it* work for $n = 2$. (Notice the similarity with the question driving my initial foray into number theory: why does it work that way?)

I systematically laid out a number of cases and looked for patterns. This was a lot of work and took a lot of time. (The process was very similar to what is outlined in the last quote on the Dedication, Inspirational Quotations page, describing the Indian mathematician Ramanujan's approach.)

As a result, although I did not successfully work out an answer to my opening question, I made the completely unexpected discovery of a method of generating what are called "Pythagorean triples" – sets of three whole numbers that satisfy the Pythagorean equation (e.g., 3, 4, and 5).

Unlike my "reinvention" of standard algebraic equations, I inductively discovered my Pythagorean triples method first, and then I eventually figured out how to prove it deductively. Needless to say, this put quite a bit of wind back in my sails!

Introduction: Confessions of a Would-Be Mathematician

At about the same time I did this, I discovered George Polya's books, and his discussions of "methods of guessing at mathematical truths and solutions" seemed to speak directly to me and to what I was doing – indeed, had been doing for many years. He said that a truly creative mathematician is a good guesser first and a good "prover" afterward.

That is certainly the *pattern* of my process of discovery – very much the way my mind works. My "psycho-epistemology," as Ayn Rand and Barbara Branden would have called it.

And so far, it seems that my method is an original, distinct alternative to the standard method for generating triples, the method that has existed for about 3,500 years. In some ways, my method works better. For more details, see Chapter 2, "A New Procedure for Generating Pythagorean Triples" and for even more details, see the Appendix: Studies on the Pythagorean Theorem.

At the time I wrote the first draft of these "confessions" in 1992, I had also just finished a first draft of what is now Chapter 2. I sent out a few review copies, and I received some welcome, encouraging feedback. I was convinced then, as now, that, if I had not unknowingly "reinvented the wheel," my discovery was a significant contribution to mathematics.

Sadly, however, thanks to numerous fruitless struggles with mathematical text-writing programs, and the high unlikelihood that my manuscript would get past the "gatekeepers" of professional journals, I decided instead to post on my web site an HTML version using fractional superscripts in lieu of the much more respectable and intelligible looking radicals for square and other roots.

Now that self-publication has come of age, I have an additional way to promote this breakthrough idea. Thus, this e-book publication, like its previous incarnation on my web site, is my way, as a non-credentialed, amateur mathematician, of putting my Pythagorean triples discovery on record.

(I am sad to report that radicals and complex fractions, which look fantastic in MS Word 2010 files and when converted to HTML and PDF files, unfortunately do *not* convert in readable fashion for Kindle publication.

My deep apologies for the unwieldy looking fractions and fractional exponents, especially in the Appendix. Email me, and I'll send you a PDF on request.)

Another discovery or insight I have hit upon more recently occurred about 18 years ago during a seminar held over the internet by the Institute for Objectivist Studies (later The Objectivist Center, presently The Atlas Society). Sometime during the fall of the 1995-96 academic year, the question was posed: what do zero exponents refer to in reality?

Even the mathematics majors in the seminar seemed stumped. But it occurred to me that, like so many concepts in mathematics, exponents don't refer to numbers of things in the world, but instead specify a *mental operation* that we do *about* numbers.

Beginning with the unit 1, I reasoned that if an exponent of 2 (i.e., some number squared) meant that the number 1 is to be multiplied by that number twice, then a zero exponent (i.e., some number to the zero power) meant that the number 1 is to be multiplied by that number *no times*. In other words, any number to the zero power is 1.

Since then, I have written up this insight, along with other comments I hope will be helpful, and I present them here in Chapter 3, along with related material on fractional exponents.

More recently, I was interested to hear Leonard Peikoff, in his 2002 lectures on induction, glowingly praise David Harriman for his interpretation of exponents in general, as a way of graphically specifying how many times a particular operation is to be performed. This is not particularly earth-shattering, since it is basically what I was taught in high school algebra.

The particular innovation I introduced in my 1995 piece to the Objectivist Center cyber seminar was to suggest that the unit 1 should be understood as the fundamental factor in *any* exponential expression. Granted that it is redundant in all other cases, it happens to be particularly clarifying in cases where the exponent is zero.

Introduction: Confessions of a Would-Be Mathematician

This brings me to my most recent insight, which I am still trying to fully understand, and about which I occasionally engage in intellectual arm-wrestling with those who find it beyond the pale. The idea is that not only is zero a "stopper" to the factoring process in exponents (5^0 is 1 *not multiplied* by 5 *any times*), it is an operation-stopper *in general*.

It's already well known that division by zero is absolutely prohibited. It's also well known that when you surreptitiously introduce zero into the denominator of a fraction, you always derive a contradictory answer. My high school freshman algebra teacher delighted in the stunned looks from our class when he produced this result.

Here is a typical example, illustrating this point. Given that a = b, then, multiplying each side by *a*, you get: $a^2 = ab$, which means that $a^2 - b^2 = ab - b^2$, which means that $(a + b)(a - b) = b(a - b)$, which means that $(a + b) = b$, which means that $a + a = a$, which means that 2a = a, which means that 2 = 1!!!

Notice that in the fourth step, the factor $(a - b)$ has been divided out from both sides of the equation. But since it is given that $a = b$, this means that $(a - b) = 0$, and thus the fourth step sneakily does the forbidden operation of division by zero. Garbage in, garbage out!

It's not difficult to see how zero is an "operation-stopper" for exponents and division. I am firmly convinced, however, that this insight applies to *every* mathematical operation, and I discuss further examples in Chapter 7.

The principle is perhaps most easy to see in addition. For instance, 5 + 0 means *not* that you *add zero* to 5, but that you *do not add anything to 5*. You do not perform the operation of addition, period. You introduce five items, and that's it. Similarly, 0 + 5 means *not* that you *add 5 to zero*, because you do not have anything to add 5 to. Instead, you introduce 5 items, and that's it.

The same reasoning applies to the case of 0 + 0. The fact that the "answer" to this is 0 does *not* mean that you *add* zero to zero to *get* zero. Instead, you do not introduce anything, and so there is not anything to start with, to which to not add anything! So again, there is not any operation to perform, and not anything on which to perform it. Again, for further details, see Chapter 7.

One thing that has been very gratifying from posting on the internet my insight about zero exponents has been all the "fan mail" (via email) from students and teachers who appreciated having a simple way to understand or explain the idea. Their questions and suggestions helped me to further refine the idea and to expand it into explaining fractional exponents as well.

I enjoy this kind of interaction, and I more generally enjoy tutoring algebra and geometry students, because they usually seem to understand math better from the general kind of approach I use. Because of this, I briefly considered about 12 years ago the possibility of changing careers from music to high school math teaching.

As a first step in this direction, I took and easily passed a general qualifying exam (C-BEST) and began preparing for three more specific qualifying exams. The material covered by the exams was so vast and my free time so limited, however, I decided not to pursue the certification in California. I am still mulling over whether to pursue substitute math teaching here in Tennessee.

Unabated, however, has been my delight in studying philosophy of mathematics and number theory and exploring at my leisure this fascinating world. Along the way, I have discovered the lectures by Pat Corvini on the nature of number, delivered under the auspices of the Ayn Rand Institute, and I most highly recommend them to anyone interested in this subject.

Will I ever be acknowledged as having blazed any new trails in mathematics? It remains to be seen, although it's arguable that my Pythagorean triples and zero exponent insights qualify as path-breaking.

But whether my contributions to mathematics "find a place in this world, or merely belong," the wonder to me is that I've done all this with little more than an obsession for math and a dogged curiosity, and certainly with none of the arcane formulas of 20th century mathematics to guide (or more likely, hamper) me.

Or, in the words of a blurb on the back of Ogilvy and Anderson's *Excursions in Number Theory*:

> It [the theory of numbers] is also a popular topic among amateur mathematicians (who have made many contributions to the field) because of its accessibility: it does not require knowledge of higher mathematics...No special training is needed – just high school mathematics, a fondness for figures and an inquisitive mind.

You rang? I must confess – and that's **Confession #3** – that I am very encouraged!

* * *

Enthusiastic recommendations:

Pat Corvini is an electrical engineer (b. ca. 1949), with a Ph.D. in Electrical Engineering from University of California at Santa Barbara (1995). She teaches mathematics at the Van Damm Academy in southern Orange County, California. As recently as 2008, she was reportedly working on a book on mathematical concepts, but I have seen no further signs of her work since then except for an announcement of an optional course on limits and universality in mathematics which she delivered to the 2009 Objectivism Conference. Her four lecture series currently available can be purchased as MP3 downloads for an extremely nominal charge from estore.aynrand.org.

_____. 2004. *The Crisis of Principles in Greek Mathematics*.

_____. 2005. *Aristotle, the Tortoise, and the Objectivity of Mathematics*.

_____. 2007. *Two, Three, Four, and All That*.

_____. 2008. *Two, Three, Four, and All That, the Sequel*.

George Polya was a Hungarian-born immigrant (1887-1985), who taught at Stanford University. One of his books is more generally accessible, while the other two listed here are more technical.

_____. 1945. *How to Solve It: a New Aspect of Mathematical Method*. Princeton University Press.

_____. 1954. *Mathematics and Plausible Reasoning – A Guide to the Art of Plausible Reasoning.* Volume 1 is entitled "Induction and Analogy in Mathematics," and Volume 2 is entitled "Patterns of Plausible Inference."

_____. 1962. *Mathematical Discovery – On Understanding, Learning, and Teaching Problem Solving.* John Wiley & Sons. This is two volumes with the same title, now available in a single paperback edition. It focuses on doing problems and then thinking about the means and methods you use to do them.

Srinivasa Ramanujan is perhaps the greatest natural mathematical talent of all time. Here are a few articles and a book about him:

_____. Borwein, Jonathan M. and Borwein, Peter B. 1988. Ramanujan and pi. *Scientific American*, Vol. 258 (February), 112-17.

_____. Kanigel, Robert. 1991. *The Man Who Knew Infinity: A Life of the Genius Ramanujan.* New York: Washington Square Press.

_____. Kolata, Gina. 1987. Remembering a "magical genius." *Science*, Vol 236 (June 19), 1519-20.

_____. Newman, James R. 1956. Srinivasa Ramanujan. In Volume I of *The World of Mathematics* (ed. Newman). New York: Dover Publications, 361-8.

_____. Peterson, Ivars. 1987. The formula man. *Science News*, Vol. 131 (April 25), 266-7.

Finally, here are some fun and interesting books to read on the subject of zero:

_____. Kaplan, Robert. 2000. *The Nothing That Is: A Natural History of Zero.* Oxford University Press.

_____. Seife, Charles. 2000. *Zero: The Biography of a Dangerous Idea.* Viking.

Chapter 1: "Speed Math"

Short-cut methods are based upon the principle of changing "difficult" numbers and processes into easier ones. [Gerard W. Kelly, *Short-Cut Math*, p. 9]

"Speed math" sometimes refers to being able to do arithmetic very quickly, as in: how fast can you add this column of figures. That's not what this essay is about.

What I mean by "speed math" is sometimes also called "short-cut math." It refers not to doing arithmetic with maximum speed in the *standard* way, but in finding non-standard ways to do calculations that allow you to do them more quickly and easily "in your head" than you can with pencil and paper or calculator.

I'm not claiming to have *invented* "speed math," but I did *discover* it on my own. This was during my senior year of high school – in the spring of 1966, to be exact.

How I discovered "speed math"

I wasn't even *looking for* "speed math." My initial curiosity had to do with doubles and squares. I knew that 2 times 2 and 2^2 were both equal to 4, but that there was no other whole number whose double and square are equal to the same number. Why was that? I wondered.

Also, as I examined a succession of cases, I saw that the gap between a numbers double and square widens steadily for larger and larger whole numbers. Again, why? Is there a pattern that characterizes this widening? A principle that explains it?

First, I set up three columns (see A, B, and C below) and started comparing the figures. Immediately I noticed that by adding the numbers in columns B and C for a given line (see column D), and then adding one more (see column E), you got the number in column B for the next line — in other words, that $x^2 + 2x + 1 = (x + 1)^2$. (Compare columns E and G.)

A	B	C	D	E	F	G
x	x^2	$2x$	$x^2 + 2x$	$x^2 + 2x + 1$	$(x + 1)$	$(x + 1)^2$
1	1	2	3	4	2	4
2	4	4	8	9	3	9
3	9	6	15	16	4	16
4	16	8	24	25	5	25
5	25	10	35	36	6	36
6	36	12	48	49	7	49

The fact that the expression at the top of column G is equal to the expression at the top of column E is something every first-year algebra student knows. It is the binomial expansion of $(x + 1)^2$, which is $(x + 1)(x + 1) = x^2 + 2x + 1$.

(There is a very simple way to derive this expansion. It's called the FOIL method for multiplying binomial terms. In this case, you multiply the First terms x and x, then the Outer terms x and 1, then the Inner terms 1 and x, and finally the Last terms 1 and 1. As a result, you get $x^2 + x + x + 1$, which simplifies to $x^2 + 2x + 1$.)

It seemed that I had just found a rather tedious way to re-invent the wheel of algebraic multiplication. What value was there in this knowledge?

Well, it occurred to me (probably because of the visual array of the various columns of numbers) that if you know the square of one number — say, 20, where $20^2 = 400$ — you can easily find the square of the next higher number — in this case, 21 — by a process of addition: $21^2 = 20^2 + 2(20)(1) + 1^2 = 400 + 40 + 1 = 441$.

This is so easy, even you and the Geico caveman can do it in your heads! Here's another example: $101^2 = 100^2 + 2(100)(1) + 1^2 = 10{,}000 + 200 + 1 = 10{,}201$. Simplicity itself!

Now, the pathway by which I arrived at this method of mental multiplication is known as *induction*. I generalized from a number of particular cases and expressed it by an equation that fits all of them.

I could instead have arrived at this equation much more quickly by simply expanding $(x + 1)^2$, as is taught to us in first-year algebra – either horizontally, using the FOIL method as I did above, or vertically, as follows:

$$
\begin{array}{r}
x + 1 \\
x + 1 \\
\hline
x + 1 \\
x^2 + x + 0 \\
\hline
x^2 + 2x + 1
\end{array}
$$

This might have been sufficient for some people to see a practical application for it. In other words, those who are better abstract thinkers than I might well have grasped this equation's usefulness (for "speed math") without needing to first go through the process of induction I carried out above.

However, it's not likely that most people could see its applicability as a shortcut for multiplication without having examined several particular instances – i.e., to have selected several values for x and plugged them into the equation.

Say, $x = 10$: $11^2 = (10 + 1)^2 = 10^2 + 2(10) + 1 = 100 + 20 + 1 = 121$.

Or, say, $x = 25$: $26^2 = (25 + 1)^2 = 25^2 + 2(25) + 1 = 625 + 50 + 1 = 676$.

In my own case, despite all my appreciation of abstract thinking, I have always found it a great help to have all the relationships displayed visually, in columns and rows of numbers. In this way, they can be viewed as concretes "out there" in reality, as examples of a single principle. Then, it is easy to further solidify the insight by following up with additional particular cases.

Getting a *feel* for the fact that this works requires a process of deduction: applying a general principle (equation, in this case) to specific instances. You *see* that it must work by generalizing from an appropriate sample of cases, and then you *verify* that it works by testing a number of other cases and *validate* it by deducing it from a more general principle you already know.

Now, we can generalize from the above equation and get a broader speed math technique that extends its usefulness. Suppose we want to add not 1, but any number, a, to x and square that sum. All we need do is square x, add twice the product of x and a, and add the square of a. In other words, $(x + a)^2 = x^2 + 2ax + a^2$.

For instance, $102^2 = 100^2 + 2\,(100)\,(2) + 2^2 = 10{,}000 + 400 + 4 = 10{,}404$.

As an exercise, I invite the reader to derive this equation both inductively, by inspection from a display of rows and columns of numbers like the one above – and deductively, by expanding the expression $(x + a)^2$.

Another useful equation is $(x - 1)^2 = x^2 - 2x + 1$. Grasping this inductively requires a similar insight to that discussed above. An example of its usefulness: $99^2 = (100 - 1)^2 = 100^2 - 2\,(100) + 1 = 10{,}000 - 200 + 1 = 9{,}801$.

And generalizing, we get $(x - a)^2 = x^2 - 2xa + a^2$. For example, $97^2 = (100 - 3)^2 = 100^2 - 2\,(100)\,(3) + 3^2 = 10{,}000 - 600 + 9 = 9{,}409$.

Again, it is instructive to derive these equations both inductively by inspecting number arrays and deductively by expansion of $(x - 1)^2$ and $(x - a)^2$.

Yet another interesting avenue to explore is to compare squares with the products of numbers equally less than and more than the squared numbers.

For instance, because we know that $100^2 = 10{,}000$, we can also see that:

1) $99\,(101) = 9{,}999 = 10{,}000 - 1 = 100^2 - 1^2$.
2) $98\,(102) = 9{,}996 = 10{,}000 - 4 = 100^2 - 2^2$.
3) $97\,(103) = 9{,}991 = 10{,}000 - 9 = 100^2 - 3^2$.

In general, $(100 - a)\,(100 + a) = 100^2 - a^2$.

Even more generally, $(x - a)\,(x + a) = x^2 - a^2$.

This equation has two useful applications.

1) First, you can quickly determine, for instance, that 23 (27) = (25 − 2) (25 + 2) = $25^2 - 2^2$ = 625 − 4 = 621.

2) Secondly, you can turn the equation around: $x^2 = (x - a)(x + a) + a^2$ and determine things like 95^2 = (95 − 5) (95 + 5) + 5^2 = 90 (100) + 25 = 9,025.

Cool, huh! Once again, I advise the reader to derive the equation inductively by inspecting number arrays and deductively by expanding the expression $(x - a)(x + a)$.

Deriving each of these equations deductively is an integral part of the standard high school algebra course, and they follow from a straightforward adaptation of the rules for long multiplication. But it takes induction or higher-level, speculative deduction to reveal their usefulness for multiplication, and neither process is given a sufficient focus in math instruction.

That is, an important part of the practical usefulness of algebra − facilitating arithmetic operations − is simply overlooked, because students are not encouraged to explore for (via induction) or to speculate about (via higher-level deduction) the results they mechanically derive.

Further references:

Many more "speed math" techniques can be found in *Short-Cut Math* by Gerard W. Kelly (Sterling, 1969; Dover repub., 1984), as well as in *The Secrets of Mental Math* by Arthur T. Benjamin (The Great Courses, 2011).

The importance of inductive "exploration" is well discussed by George Polya in *Induction and Analogy in Mathematics* (Princeton University Press, 1954).

Postscript − sums of counting numbers: Returning to "the scene of the crime" in 2012, I discovered some more interesting results by rather simple arithmetic comparison of columns of numbers. I started tinkering one day with sums of counting numbers.

The first sum is simply 1, the second is 1 + 2 = 3, the third is 1 + 2 + 3 = 6, and so on. The nth sum, it turned out, is $(n^2 + n)/2$.

The "speed math" in this – the practical application, if you could call it that – is that if you want to know the sum of the first n counting numbers, just divide the sum of the largest one and the largest one's square by 2.

In retrospect, I saw that I could have arrived at the formula for the sum of the first n counting numbers by making three columns showing the sum for each 1 through n, the square for each n, and the sum of n and its square. From this, the answer would have jumped out at me by simply comparing the sum of 1 through n with $n^2 + n$, the former being ½ of the latter.

n	sum of 1 through n	n^2	$n^2 + n$
1	1	1	2
2	3	4	6
3	6	9	12
4	10	16	20
5	15	25	30

However, I had no reason to construct this particular table of numbers. I wasn't curious about the squares of the counting numbers, just the sums of them. So, I proceeded in a much more intuitive, exploratory way, as illustrated below. I summed the counting numbers 1 through n for each counting number n. Then I found the lowest two factors for each sum.

CN1: 1   sum of CN's: 1 = 1 (1)

CN2: 1 + 2   sum of CN's: 3 = 1 (3) = (2/2) (2 + 1)

CN3: 1 + 2 + 3   sum of CN's: 6 = 2 (3)

CN4: 1 + 2 + 3 + 4   sum of CN's: 10 = 2 (5) = (4/2) (4 + 1)

CN5: 1 + 2 + 3 + 4 + 5   sum of CN's: 15 = 3 (5)

CN6: 1 + 2 + 3 + 4 + 5 + 6   sum of CN's: 21 = 3 (7) = (6/2) (6 + 1)

CN7: 1 + 2 + 3 + 4 + 5 + 6 + 7   sum of CN's: 28 = 4 (7)

CN8: 1 + 2 + 3 + 4 + 5 + 6 + 7 + 8   sum of CN's: 36 = 4 (9) = (8/2) (8 + 1)

CN9: 1 + 2 + 3 + 4 + 5 + 6 + 7 + 8 + 9   sum of CN's: 45 = 5 (9)

Chapter 1 – Speed Math

*CN*10: 1 + 2 + 3 + 4 + 5 + 6 + 7 + 8 + 9 + 10   *sum of CN's*: 55 = 5 (11) = (10/2) (10 + 1)

*CN*11: 1 + 2 + 3 + 4 + 5 + 6 + 7 + 8 + 9 + 10 + 11   *sum of CN's*: 66 = 6 (11)

I then observed an interesting pattern in the factors. The first factor was 1 twice, 2 twice, 3 twice, etc., a repetition of each counting number. The second factor was 1 once, then 3 twice, 5 twice, 7 twice, etc., a repetition of each odd number.

Based on this, I decided to split up the cases and examine first what was happening with the even numbers. I wanted to see if I could find a consistent pattern that could be expressed algebraically – and then do the same for the odd numbers. I immediately noticed that the factors of the sums for each even number were half of that number multiplied by one more than that number.

For instance, for 1 through 10, the sum 55 has factors 5 and 11, and these are ½ of 10 and one more than 10, respectively. In general, for the even numbers, the sum is expressible as $(n/2)(n+1) = (n^2 + n)/2$.

For the odd numbers, the situation was a little more convoluted, but not greatly so. The factors of the sums for each odd number were one more than half of the previous number multiplied by that number.

For instance, for 1 through 9, the sum 45 has factors 5 and 9, and these are one more than ½ of 8 and one more than 8. In general, for the odd numbers, the sum is expressible as $([(n-1)/2] + 1) n = [(n+1)/2] n = (n^2 + n)/2$.

Combining the results, I saw that, for both odd and even counting numbers, the sum of the first *n* counting numbers is always $(n^2 + n)/2$.

In an Internet search, I was not able to find this exact discovery, but I did find something closely related that was referred to as "Floyd's Triangle." This diagram sums *rows* of counting numbers arrayed in a Christmas tree-like pattern, one descending rows containing 1, then 2 and 3, then 4 and 5 and 6, etc. (It is used in computer science education, and it was named after Robert Floyd.)

How the Martians Discovered Algebra

The Wikipedia article on Floyd's Triangle pointed out that the nth row sums to $n(n^2 + 1)/2$, which is $(n^3 + n)/2$. Deriving this inductively in the manner I did above would perhaps be an interesting project. I leave this as an exercise for the motivated reader.

Postscript #2 – what is the point of speed math? In his review of this book (*Journal of Ayn Rand Studies*, December 2014), Fred Seddon pointed out that various shortcut techniques described in this chapter can be done more quickly by calculator. True, I did misspeak myself when I said that "speed math" was faster than using a calculator. But what about when your calculator's battery dies or the electricity goes out? Or you're driving somewhere and you want to do a calculation and you don't have a hands-free calculator (if there are such things)? Knowing how to tie a Windsor knot may be a needless skill for those who are perfectly happy with a clip-on (or no) tie – and being able to change a flat tire may be useless know-how for those who have AAA. Speed math falls somewhere along that spectrum of necessity and convenience.

Dr. Seddon also noted that he rarely needs to find the square of 99 (which was just *one* of my examples), but you don't have to look very far to find practical examples that might come up in real life. Suppose you want to estimate how much it will take to feed 25 people at $15 per serving. If you know the formula $(x + y)(x - y) = x^2 - y^2$, you can quickly see that it will cost $(20^2 - 5^2)$ or $(400 - 25)$ or $375. Similarly, feeding 27 people at $13 per serving comes out to $351. But there are numerous other scenarios in which speed math, while technically possible, is too mentally difficult to be helpful – such as figuring out the cost of feeding 28 people at $14 per serving. In each of these three cases, you first quickly try to see if the number halfway between the two numbers is one for which you know the square. Many people know the square of 20 (which is 400), so any pair of numbers equidistant from 20 can be quickly multiplied using the formula. Most people do *not* know the square of 21 (which is 441), so a pair of numbers equidistant from it (such as 28 and 14) *cannot* be quickly multiplied using the formula. The moral of the story is: speed math is an opportunistic tool. If you know the tool and can quickly see whether it's useful or not, you can do the operation you want to do quickly, without having to reach for another device or application to do it for you.

Chapter 2: A New Procedure for Deriving Pythagorean Triples

Pythagorean triples offer an almost unlimited plethora of opportunities for finding amazing numerical relationships, which the reader is encouraged to pursue.
[Alfred S. Posamentier, *The Pythagorean Theorem: the Story of Its Power and Beauty*, p. 167]

Sometime during the early part of 1988, I read an article in the *Orange County Register*, "Number Theorist Might Have Proof to Theorem of 1637." This referred to a claim made by the French mathematician, Pierre de Fermat. The claim, known as Fermat's Last Theorem, says that there is no whole number n, larger than 2, for which $x^2 + y^2 = z^2$.

In other words, Fermat claimed, the Pythagorean Theorem, $x^2 + y^2 = z^2$, does have solutions – *infinitely many*, as a matter of fact – but there is no higher order equation that does, whether $x^3 + y^3 = z^3$, $x^9 + y^9 = z^9$, $x^{100} + y^{100} = z^{100}$, or whatever whole number you might choose for n.

Fermat said he had a proof for this claim, but either he never wrote it down, or it was lost. Recently, a Japanese number theorist, Yoichi Miyoaka, said he might have a proof, which he outlined at the Max Planck Institute in Bonn, Germany early in 1988.

Had this proof been genuine, it would have closed the door on a 350-year-old mystery. As it later turned out, however, Miyoaka's proof was not valid.

In the meantime, I got curious about the issue myself. What is it about higher powers that won't let the equation work like it does for $n = 2$? What is it about $n = 2$ that *lets* the equation work?

Could I find clues to this mystery by making a deep, careful study of the Pythagorean Theorem – and then somehow try to generalize? And what else might I find out about the Pythagorean Theorem in the process? I decided it would be worth looking into.

The following is a highly condensed version of work done by me between early 1988 and October 1992. For the complete version, showing all of the various inductive processes involved, see the Appendix to this book.

Part I: Derivation of a new procedure for generating Pythagorean triples

The following is the derivation of a new procedure for generating Pythagorean triples – i.e., solutions of the Pythagorean equation that have integral values:

Step 1:

$$x^2 + (x + a)^2 = (x + b)^2$$
$$x^2 + x^2 + 2ax + a^2 = x^2 + 2bx + b^2$$
$$x^2 + 2ax - 2bx + a^2 - b^2 = 0$$
$$x^2 + 2x(a - b) + (a^2 - b^2) = 0$$
$$x^2 + 2x(a - b) + (a - b)^2 - (a - b)^2 + (a^2 - b^2) = 0$$
$$[x + (a - b)]^2 - (a - b)^2 + (a^2 - b^2) = 0$$
$$[x + (a - b)]^2 - a^2 + 2ab - b^2 + a^2 - b^2 = 0$$
$$[x + (a - b)]^2 + 2b(a - b) = 0$$
$$[x + (a - b)]^2 - 2b(b - a) = 0$$
$$[x + (a - b)]^2 = 2b(b - a)$$
$$x + (a - b) = [2b(b - a)]^{1/2}$$
$$x = [2b(b - a)]^{1/2} - (a - b)$$

Result 1: $x = [2b(b - a)]^{1/2} + (b - a)$

Step 2:

Since x, $x + a$, and $x + b$ are all integers, a and b must be integers as well. Therefore, $a + b$ must also be an integer, and thus there must be some rational number $c = x/(a + b)$, or:

Result 2: $x = c(a + b)$

Then, equating the right sides of (1) and (2), it follows that:

$$c(a + b) = [2b(b - a)]^{1/2} + (b - a)$$
$$c(a + b) - (b - a) = [2b(b - a)]^{1/2}$$
$$c^2(a + b)^2 - 2c(a + b)(b - a) + (b - a)^2 = 2b^2 - 2ab$$

Chapter 2 – A New Procedure for Deriving Pythagorean Triples

$a^2c^2 + 2abc^2 + b^2c^2 + 2a^2c - 2b^2c + b^2 - 2ab + a^2 = 2b^2 - 2ab$
$a^2c^2 + 2a^2c - 2b^2c + 2abc^2 + b^2c^2 + b^2 = 2b^2 - a^2$
$c^2(a^2 + 2ab + b^2) + (2c + 1)(a^2 - b^2) = 0$
$c^2(a+b) + (2c+1)(a-b) = 0$
$ac^2 + bc^2 + 2ac - 2bc + a - b = 0$
$(a/b)c^2 + c^2 + 2(a/b)c - 2c + a/b - 1 = 0$
$(a/b)(c^2 + 2c + 1) = -c^2 + 2c + 1$
$(a/b)(c+1)^2 = -(c^2 - 2c - 1)$
$(a/b)(c+1)^2 = -(c^2 - 2c + 1 - 2)$
$(a/b)(c+1)^2 = 2 - (c^2 - 2c + 1)$
$(a/b)(c+1)^2 = 2 - (c-1)^2$

Result 3: $a/b = [2 - (c-1)^2]/(c+1)^2$

Step 3:

Now select any rational number c, where $c < -1$ or $0 < c$. Solve (3) for a/b. Set the numerator of a/b equal to a, and the denominator equal to b. Solve (1) or (2) for x, which yields the values for $x + a$ and for $x + b$. **This is the procedure in its entirety.**

Note 1: taking $-1 < c < 0$ does not yield Pythagorean triples. At least one of x, $x + a$, or $x + b$ is 0 or negative.

Note 2: for $0 < c \leq (2^{1/2} + 1)$, $x < x + a < x + b$, and for $c < -1$ and $(2^{1/2} + 1) < c$, $x + a < x < x + b$.

Note 3: c must be rational. Assume the contrary. It follows that $x/c = (a + b)$ is irrational. From this, it follows that a or b or both are irrational. From this, it follows that $(x + a)$ or $(x + b)$ or both are irrational.

Note 4: by substitution into (2), the following limits obtain:

> When $c \to$ infinity, the limit of $x/(x+b)$ is 4/5, the limit of $(x+a)/(x+b)$ is 3/5, and the limit of $x/(x+a)$ is 4/3.

> When $c \to 0$, the limit of $(x+a)/(x+b)$ is 1, the limit of $x/(x+a)$ is 0, and the limit of $x/(x+b)$ is 0.

> When $c = 2^{1/2} + 1$, $a = 0$, $(x+a)/x = 1$, $(x+b)/x = 2^{1/2}$, and $(x+b)/(x+a) = 2^{1/2}$. Thus, there can be no isosceles right triangle with all sides of integral length.

How the Martians Discovered Algebra

Part II: Comparison of the author's method with the traditional procedure for generating Pythagorean Triples

As Kline notes in *Mathematical Thought from Ancient to Modern Times* (Oxford, 1972, p. 31):

> The Pythagoreans devised a rule for finding triples of integers which could be the sides of a right triangle...They found that when m is odd, then m, $(m^2 - 1)/2$, and $(m^2 + 1)/2$ are such a triple. However, this rule gives only some sets of such triples. Any set of three integers which can be the sides of a right triangle is now called a Pythagorean triple.

In *Excursions in Number Theory* (New York: Dover, 1966, pp. 66-67), Ogilvy and Anderson give this derivation of the standard method of generating Pythagorean triples, which dates back to ancient Greek and Babylonian times (c. 500 BC):

> To guarantee that we get new triangles [as opposed to linear multiple ones] we can consider only the primitive solutions, meaning x, y, z having no factor in common....x and y cannot both be odd [according to congruence theory]. On the other hand, x and y cannot both be even, for then z would be even and the solution would not be primitive. Suppose, then, that x is odd and y is even. This means that we can write $x^2 + 4u^2 = z^2$, with no two of x, u, z having a common factor. Therefore, $4u^2 = z^2 - x^2 = (z + x)/(z - x)$. But x and z are odd. Hence, both $(z + x)$ and $(z - x)$ are even, say $z + x = 2s$, $z - x = 2r$. That is, $4u^2 = 2s2r$, or $u^2 = rs$. Now the two equations in z and x yield, on addition, $z = r + s$, and on subtraction, $x = s - r$. If r and s had a common factor, z and x would have it also. But z and x are relatively prime; therefore so are r and s. As a consequence, the equation $u^2 = rs$ requires that r and s are each perfect squares (neither can pick up a "matching" factor from the other), say, $r = n^2$, and $s = m^2$. Then $u = mn$, and we have finally:
>
> $x = m^2 - n^2$
>
> $y = 2mn$
>
> $z = m^2 + n^2$
>
> ...[W]hat are the restrictions on m and n? First, $m > n$, so that x will be positive. Second, m, n must have no common factor, or it

would be shared (twice) by x, y, and z. Third, m and n must not both be odd, for then x and z would share the factor 2.

There is a (relatively) simple relationship between the "inputs" of the traditional method of generating Pythagorean triples and my method presented above. The derivation of the "transformation equations" (i.e., the equations that link the two methods) is as follows:

Given x odd and (m, n) restricted as above, $a = 2mn - x = 2mn - m^2 + n^2$, and $b = m^2 + n^2 - x = 2n^2$. So, since $x = c(a + b)$, $m^2 - n^2 = c(2mn - m^2 + n^2 + 2n^2)$, and $\boldsymbol{c = (m - n)/(3n - m)}$. If x is even, $c = (g + h)/(g - h) = n/(m - 2n)$, so that $\boldsymbol{c = n/(m - 2n)}$.

Given x odd and rational number $c = g/h$, where $0 < c$, or $c < -1$, $g/h = (m - n)/(3n - m)$, $mh - nh = 3ng - mg$, $m(h + g) = n(3g + h)$, and $\boldsymbol{m/n = (3g + h)/(g + h)}$. Set $m = 3g + h$ and $n = g + h$. If x is even, $m/n = [3(g + h) + (g - h)]/[(g + h) + (g - h)] = (4g + 2h)/2g$, so that $\boldsymbol{m/n = (2g + h)/g}$.

For more information on formulas for generating Pythagorean triples, see:

http://www.maths.surrey.ac.uk/hosted-sites/R.Knott/Pythag/pythag.html

http://en.wikipedia.org/wiki/Pythagorean_triple

http://en.wikipedia.org/wiki/Formulas_for_generating_Pythagorean_triples

http://www.maths.surrey.ac.uk/hosted-sites/R.Knott/Fibonacci/fibmaths.html#pythagfib

Posamentier, Alfred S. *The Pythagorean Theorem: The Story of Its Power and Beauty*. Amherst, New York: Prometheus Books, 2010.

Chapter 3: What's the Deal with X-to-the-Zero Power? A Brief Note on the Non-Fiction of Zero Exponents

If you try to philosophize your way around this question you get caught up in terrible verbal cat's-cradles...We want to find out what 0^0 power is by slipping stealthily toward it. [Robert Kaplan, *The Nothing That Is: A Natural History of Zero*, pp. 117, 165]

One of the big stumbling blocks for high school algebra students – and, it appears, some professional logicians! – is the issue of zero and negative exponents. What do those little numbers mean that we put above and to the right of numbers and letters? What, if anything, in reality do they refer to?

I hope this short piece will help to strip away some of the mystery surrounding these little beasties.

From September 1995 to summer 1996, I participated in an internet seminar on the nature of propositions. In December, there was a brief but fascinating discussion of the possible meaning of mathematical expressions such as those with zero exponents. 5^0, $(9,000,000)^0$, $(3/5)^0$ – you name it, and it's equal to 1. Why should this be so? What in reality could it possibly refer to?

Some of the participants in the seminar seemed mystified by the mathematical rule that any number to the "zero" power is equal to 1. One of them even suggested that it was a "convenient fiction" to make the math come out right.

Consider positive exponents: $5^2 = 25$, *because* (we are told by our teachers) the square (exponent of two) means that you have two factors of the number 5 (i.e. you multiply 5 times itself). Thus, $5^4 = 5 \times 5 \times 5 \times 5 = 625$.

Negative exponents are a little trickier: $5^{-2} = 1/25$, because a negative power is (supposedly) defined as the exponent of a number that is the *denominator* of a fraction with a numerator of 1. In this case, $5^{-2} = 1$ divided by 5^2, which is 1 divided by 5 x 5, which is 1/25.

But what possible meaning can there be to multiplying a number by itself *zero* times? There is no such mathematical operation, is there? And even if there were, why should the result always be 1?? Yet, when we are taught to multiply an integer times itself by adding its exponents, we *have to* do it that way so that the math will come out right.

> For instance: $6^2 \times 6^2 = 6^{2+2} = 6^4 = 1,296$, by the exponent rule, while $6^2 \times 6^2 = (6 \times 6)(6 \times 6) = 36 \times 36 = 1,296$. No problem there.

> And: $6^0 \times 6^2 = 6^{0+2} = 6^2 = 36$, by the exponent rule, while $6^0 \times 6^2 = 6^0 (6 \times 6) = 6^0 \times 36$. For the math to "come out right," 6^0 must be 1. (So far, so good?)

> And: $6^3/6^3 = 6^{3-3} = 6^0$, by the exponent rule, while $6^3/6^3 = (6 \times 6 \times 6)/(6 \times 6 \times 6) = 218/218 = 1$. Again, for the math to work properly, 6^0 must be 1.

So, what is the deal with zero exponents? They obviously work, but *why* do they work? Are they just a "convenient fiction" – something to help the math "come out right"? Or is there something that our math teachers haven't told us?

The confusion comes in the way that exponents are described. The basic function of an exponent is to relate the operations of multiplying or dividing *not* (primarily) to the factor – i.e., the number of which it is an exponent – but to *the unit number, 1*.

(This, of course, is the basis of the much-dreaded "scientific notation," which I think is how students should be taught exponents, rather than the more colloquial "times-itself" approach.)

This relation to the unit 1 is *explicit* for negative exponents: x^{-3}, for example is 1 *divided by* x^3. For positive exponents, it is left implicit (since 1 is the multiplicative identity), but is more fully stated, for example, as: x^3 is 1 *multiplied by* x^3.

Chapter 3 – What's the Deal with X-to-the-Zero Power?

In light of this, a zero exponent's non-fictional relation to the unit 1 becomes clearly plausible. We simply need to *extend the method used for negative exponents to positive and zero exponents*. It is *this* that our math teachers didn't tell us about working with exponents.

Here's how it works:

> For any real number, r, a positive exponent, n, indicates that the unit 1 is to be considered as having been multiplied by r a total of n times.
>
> Similarly, a negative exponent, $-n$, indicates that the unit 1 is considered as having been divided by r a total of n times.
>
> A *zero* exponent, by contrast, indicates that the unit 1 is considered *out of* any multiplicative or divisive relationship to the base number r. We are to consider only the unit 1 and to *refrain from* combining it (by multiplication or division) with the exponent's base number.

In addition to the simplicity and clarity this approach provides, it also applies with the same results to the special case where $x = 0$, which is precisely where the conventional view breaks down.

By convention, 0^0 is *not* "obviously" equal to 1, but instead is undefined. The conventional explanation of x to the zero power argues that: $x^0 = x^{1-1} = x^1/x^1 = 1$. But clearly this cannot be the case when $x = 0$, because $0^1/0^1 = 0/0$, which is also undefined.

By *my* perspective on zero as a non-operator, however, 0^0 in fact *does* equal 1.

(1) By definition, x^n is the unit 1 multiplied n times by the factor x.

(2) Where $n = 0$, x^0 is the unit 1 *not multiplied any times* by the factor x – i.e., $x^0 = 1$.

(3) Where $n = 0$, and $x = 0$, 0^0 is the unit 1 *not multiplied any times* by the factor 0 – i.e., $0^0 = 1$.

So, by my argument, $x^0 = 1$ for *any* number x including zero. Contrast this with the conventional argument, which says that $x^0 = 1$ for any number x

except zero, where it is undefined. By yet another argument, $x^0 = 1$ for any number x *except* zero, in which case $x^0 = 0$ (because 0 to any other power than 0 is 0, so 0^0 should be zero, too).

Which is right? I would suggest that *mine* is correct, because it gives a completely consistent result for all values of x, and it relies not on *undefined results of operations*, but merely on an operation *not being performed* on the unit 1.

In mathematics theory, this is referred to as an "empty product" or a "nullary product," which is defined as "the result of multiplying no factors" (i.e., of *not* multiplying *any* factors). See Wikipedia, "Empty product." The empty product is said to be "by convention equal to the multiplicative identity 1."

This appears to be a completely arbitrary rule – until and unless you realize the fact that the multiplicative identity, the unit 1, is everywhere present as the implied factor of every other number, including numbers with a zero exponent. Then, of course, it makes perfect sense to conclude that any number to the power 0 is the multiplicative identity 1 *not* multiplied by that number *any* times – and that this would be just as true when that number is 0.

A very interesting web site gives several other arguments why 0^0 should be equal to 1 (rather than equal to 0 or undefined). Check out <u>Zero to the Zero Power – Mudd Math Fun Facts</u>.

You will note that while the author agrees with the conventional view that 0^0 is undefined, he also "muddies" the waters by stating that, if it "should" be defined, there are a number of reasons why it "should" be equal to 1. (In particular, take a look at the fourth of the five reasons. It vaguely resembles my approach, though it is very sketchy.)

What is important to realize is that *all* power operations should be defined in terms of the unit 1 being multiplied (or not!) by some factor x a total of n times. This allows us to see that any number to the zero power is 1, without having to invoke undefined results of operations to explain away the exception, 0^0.

Chapter 3 – What's the Deal with X-to-the-Zero Power?

Now we can see the root of the confusion inherent in the way most of us are taught about zero exponents.

The exponent 0 indicates that the unit 1 should be considered not as *being* multiplied (or divided) by some base number "zero times," but *simply as itself* – i.e., as *not* having been multiplied (or divided) by the base number *any* times.

And for all other exponents, the unit 1 should be considered as the base number that *is* multiplied or divided by the number with the exponent, the number of times equal to its exponent.

All of this reminds us of something very important about the nature of numbers in general. *The number 1 is an implied factor of all mathematical expressions* – e.g., $(x + y)(x - y) = 1 (x + y)(x - y)$ – and the zero exponent is simply another way in which this omnipresence of the unit factor 1 is acknowledged.

Postscript – comparing this perspective to the standard approach: To any professional mathematicians who may be reading this, I apologize for the inelegance of the preceding explanation. And to any epistemologists or logicians who are tuned in, I apologize for the absence of hierarchical development of this topic and of genus-differentia definitions of my terms.

For the general reader, however, I'm sure that the foregoing suffices to show clearly that there really is a crucial difference in how zero exponents are explained by the standard theory and by my model.

From the standard perspective, in dealing with explaining the meaning of exponents, you have to say that any number x to some power n is that number x multiplied by itself n times, unless the power is 0, in which case x^n is 1.

From my perspective, all I have to say is: any number x to some power n is the unit 1 taken times x n times, unless n is 0, in which case the unit 1 is *not* taken time x *any* times.

For my money, my unit-factor model is much more elegant and integrated. It's single-premised, and it appears less arbitrary.

First of all, where did the answer "1" come from, on the standard perspective? You have to do a *lot* of manipulation to show it.

Also, think of the analogy between my unit-factor foundation for exponents and scientific notation – e.g., $100 = 1.0 \times 10^2$, and $3,952 = 3.952 \times 10^3$.

We already have the precedent of analyzing multiple-digit numbers in terms of a single-digit (with decimals) number multiplied by factors of 10. So, why not address both issues by the same approach? Why not analyze power numbers in terms of a single-digit number (1) multiplied (or not!) by factors of the number that is being powered?

If I were King of All Mathematics, I would institute this way of conceptualizing powers immediately! It might forestall a lot of perplexity in young math students. In fact, I have gotten quite a number of grateful emails from high school and college students who understood zero powers, negative powers, and fractional powers for the first time from my online essays.

I received the following interesting email in 2004 from someone who read an earlier version of this piece on my web site.

> I just read your explanation of why any number to the zero power is 1. But, I didn't really follow you. Just because you can form a nice pattern, does that always mean that your definition is a good one? Or that it is consistent with the rest of the laws of arithmetic? Seems like your argument just begs this question. I have always thought that maybe we just define it to be 1 and then show that it is consistent with the laws of arithmetic. Could you provide further explanation? Thank you for your time.

I replied:

> Greetings. I'm sorry that you aren't following my argument.
>
> Let me try another approach. Consider scientific notation, where we express 500 as 5.0 times 10 to the 2nd power. We are *describing* (defining?) 500 as 5×10^2.
>
> OK, what I am doing with the exponent topic is an *analogy* to this. I said, why not describe any power as 1 multiplied by a number the power number of times?

Chapter 3 – What's the Deal with X-to-the-Zero Power?

See, I was trying to answer someone who said what does zero power refer to in reality? I realized that it didn't refer to anything in reality, but it *did* refer to a certain *operation* being done zero times. But done *to what* zero times? *To the number 1.*

Naturally, since this is a correct description of what is going on, it means that when you use it, it works, because this is what is going on. But you don't need to think of this in order to use it. All you have to do is remember that's what a zero power is, and use it consistently.

But this is true with lots of things that we learn in mathematics. We learn the basis or proof of it, then go on and automatize it and use it as a practical tool.

However, I don't think it is good, as you suggest, to think of $x^0 = 1$ as an *arbitrary* thing. It is not arbitrary, it is *real*. It refers to *a real operation that is not being carried out* (in relation to the unit 1).

Hope this helps.

Roger Bissell, paramathematician

Postscript 2 – extending this approach to fractional exponents: An interested reader of this material on zero exponents (which I had posted to my web site) wrote the following:

> How about fractional exponents? How can a number with a fractional exponent be expressed (in English) in reference to the unit 1 as you do for positive, negative, and zero exponents?
>
> To express the unit 1's relationship to positive and negative exponents, you say, "For any real number, r, a positive exponent, n, indicates that the unit 1 is to be considered as having been multiplied by r a total of n times. A negative exponent, $-n$, indicates that the unit 1 is considered as having been *divided* by r a total of n times."
>
> To express the unit 1's relationship to zero exponents, you say, "The unit 1 should be considered ... as not having been multiplied (or divided) by the base number any times."
>
> OK, so take x raised to the $1/n$ power. It is equal to the nth root of x. A fractional exponent – specifically, an exponent of the

form $1/n$ – means to take the nth root instead of multiplying or dividing. For example, $4^{1/3}$ is the 3rd root (cube root) of 4.

Obviously, the unit 1 is explicit here in the numerator of the fraction, but how would you rewrite the definition of a fractional exponent "an exponent of the form $1/n$ – means to take the nth root" and express this in terms of how the unit 1 is central to the operation being performed in a fashion similar to how you expressed the unit 1's centrality in the cases of positive, negative, and zero exponents?

OK, I scratched my head a bit, and I figured it out!

For any real number, r, the positive fractional exponent $1/n$ indicates that the unit 1 is to be considered as having been *multiplied by* one of r's n equal factors. The negative fractional exponent $-1/n$ indicates that the unit 1 is considered as having been *divided by* one of r's n equal factors.

For example if $r = 8$, and $n = 3$, we want to express $8^{1/3}$ (i.e., the cube root of 8) like this: the unit 1 *multiplied by* one of 3 equal factors of 8. And if $r = 256$, and $n = 4$, we want to express $256^{-1/4}$ (i.e., the negative fourth root of 256) like this: the unit 1 *divided by* one of 4 equal factors of 256.

The arithmetic is: $8^{1/3} = 1 \times 2 = 2$, and $256^{-1/4} = 1 \times 1/4 = 1/4$.

All right, let's push on. What about *mixed fraction* exponents – for instance, $8^{2/3}$ or $256^{-3/4}$? Well, the former is the unit 1 *multiplied by* one of 3 equal factors of 8, a total of 2 times – and the latter is the unit 1 *divided by* one of 4 equal factors of 256, a total of 3 times.

The arithmetic is: $8^{2/3} = 1 \times 2 \times 2 = 4$, and $256^{-3/4} = 1/4/4/4 = 1/(4 \times 4 \times 4) = 1/64$.

The principles for fractional exponents are thus seen to be analogous to those for integral exponents, and those for mixed fractional exponents are a combination of the other two.

So, to generalize for any rational exponent: for any real number, r, a positive rational exponent, m/n, indicates that the unit 1 is to be considered as having been *multiplied by* one of r's n equal factors a total of m times. A negative rational exponent, $-m/n$, indicates that the unit 1 is considered as having been *divided by* one of r's n equal factors a total of m times.

Chapter 3 – What's the Deal with X-to-the-Zero Power?

Postscript 3 – further clarifications: I received two interesting emails regarding this material on fractional exponents, which I had posted on my web site in 2002. One was undated in my email folder, but the sender simply said:

> I've a passing interest in fractional exponents. I saw your article, which showed examples of $4^{1/2}$ and $8^{1/3}$. Could you please demonstrate the following two examples: $3^{1/2}$ and $7^{1/3}$? Thanks! P.S. There is a very practical point to this, or I wouldn't be asking. I'm NOT a mathematician, but I do have a bit of a fascination with properties and algorithms.

I replied:

> OK. On my web page, I wrote: "To generalize for any rational exponent: For any real number, r, a positive rational exponent, m/n, indicates that the unit 1 is to be considered as having been multiplied by one of r's n equal factors a total of m times."
>
> So, taking your first requested example, $3^{1/2}$, the real number r is 3, and the rational exponent m/n is ½, which indicates that the unit 1 is to be considered as having been multiplied by one of 3's 2 equal factors a total of 1 time. That is, 1 is multiplied by the square root of 3 one time.
>
> In your second example, $7^{1/3}$, the real number r is 7, and the rational exponent m/n is 1/3, which indicates that the unit 1 is to be considered as having been multiplied by one of 7's 3 equal factors a total of one time. That is, 1 is multiplied by the cube root of 7 one time.
>
> Similarly (just for some additional examples), $3^{3/2}$ would be regarded as 1 multiplied by the square root of 3 three times, and $7^{2/3}$ would be regarded as 1 multiplied by the cube root of 7 two times.
>
> I hope it is clear that the essence of this approach is to put the unit 1 in the forefront of how we *think* about this notation. It is vital in understanding zero power numbers, but it is good practice to keep it in mind for fractional exponents.

In November of 2004, I had the following email exchanges (excerpting for relevant content):

I can't just memorize rules because I don't really understand what I'm doing and I will get confused almost immediately.

For example, I am stuck trying to understand the reciprocal rule for negative exponents. I know the rule is to take the reciprocal of the base number and change the exponent to positive but I don't understand why. It doesn't make sense to me. I figure that if 3^2 is 3 times 3, then 3^{-2} would be 3 divided by 3. Ahhhh! Help! Thank you so much.

I responded:

Try to think of it this way:

3 to the 2 power is 1 x 3 x 3 (two times)

3 to the 1 power is 1 x 3 (one time)

3 to the 0 power is 1 *not* multiplied by 3 *any* times

3 to the –1 power is 1 divided by 3 (one time)

3 to the –2 power is 1 divided by (3 x 3) (two times)

You know that minus powers have to involve a fraction. So, they have to have 1 on top. This is why it's so important to always put everything in terms of 1. Not only the zero power, but plus and minus powers, too. You start with 1 and then you do things to it, either multiply or divide.

What you do to it is dictated by the plus or minus (multiply or divide). *How many times* you do it is dictated by the value of the power. E.g., 4 power means multiply 1 by the number (whatever it is) 4 times. So, 3^4 means multiply 1 by 3 four times (= 81). E.g., -3 power means *divide* 1 by the number (whatever it is) 3 times. So, 4^{-3} means *divide* 1 by 4 three times (= 1/64).

The same way you understood what I wrote about zero powers will help you to understand negative powers, too! I hope this helps.

The questioner wrote back and said simply: "One quick question. How is it that 'I know minus powers have to involve a fraction?' Thanks."

I responded:

Just think of a fraction as a visual, vertical way of showing that you are *dividing*. For instance, 1 divided by 3 can be shown visually and vertically as 1/3. Remember, we start with 1. A plus power means multiply 1 a certain number of times by some number. A minus power means divide 1 a certain number of times by some number.

For instance, 3^{-3} means: divide 1 by 3 three times. 1 divided by 3 one time is 1/3. 1 divided by 3 two times is 1/9. 1 divided by 3 three times is 1/27. Or, 1 divided by 3 is 1/3, 1/3 divided by 3 is 1/9, and 1/9 divided by 3 is 1/27. Or, 1/3 x 1/3 x 1/3 = 1/27. Or, 1/(3 x 3 x 3) = 1/27.

Those are all different ways of showing the same thing. Remember, if you divide by 3, you are multiplying by 1/3. This is from the basic rule of multiplying fractions. Also, remember the rule about multiplying fractions that says you multiply the numerators together, and you multiply the denominators together.

That may seem like a bunch of stuff to remember, but it's mostly 4th grade arithmetic. The extra twist is how it helps to understand what plus and minus powers do. That's what you have to connect up with your knowledge of multiplying fractions. Hope this helps!

More recently, I was given the following challenge by someone who apparently figured that the largeness of the decimal fraction would melt my brain, or something. Or perhaps she was just skeptical that my approach to analyzing fractional exponents really is as effective and practical as it is.

At any rate, here is the challenge: what does the number 4,325 correspond to in reality? And what does the power 0.4325 refer to in reality?

Yes, it looks rather formidable – mind-wrenchingly tedious, at the very least. But remember, if any mathematical expression is valid, it must be derived from and ultimately reducible to something in reality or some mental procedure in regard to reality.

Also remember: in general, powers mean: begin with the unit 1, *and then* multiply (or don't multiply) it by some specified number of factors of a certain number.

So, just as we learn that 4,325 means the number of things counted by 4 groups of 1,000 things + 3 groups of 100 things + 2 groups of 10 things + 5 things, I think that the power 0.4325 can be (ultimately) unpacked in the same general way. But please don't make me go through the steps on this. My head hurts already.<whimper>

Oh, all right...here is the general principle, as developed in this essay. For any real number, r, a positive rational exponent, m/n, indicates that the unit 1 is to be considered as having been multiplied by one of r's n equal factors a total of m times.

In the present example, the positive rational exponent is 4,325/10,000. That means, for instance, that $8^{0.4325}$ is 1 multiplied by one of 8's 10,000 equal factors a total of 4,325 times. If this sounds bizarre, it is in principle no more bizarre than saying (as we learn in high school) that $8^{2/3}$ is the square of the cube root of 8 (or the cube root of the square of 8).

Oh, you'd like me to *compute* the answer? Sorry, the "right side" of my brain doesn't provide details, just method! For further details, re-read this chapter.

Chapter 4: How the Martians Discovered Algebra

> *Imagine...a species of thinking atoms; they have some kind of sensory apparatus but, given their size, no eyes or tactile organs and therefore no color or touch perception. Such creatures, let us say, perceive other atoms directly, as we do people; they perceive in some form we cannot imagine. For them, the fact that matter is atomic is not a theory reached by inference, but a self-evidency.* [Leonard Peikoff, *Objectivism: The Philosophy of Ayn Rand*, p. 43]

October 29, 2097, North American Sector, Preliminary Report

It is a little known historical fact that, early in the 20th century, Albert Einstein briefly established contact with beings from the Red Planet and learned of the extremely odd development of their scientific and mathematical knowledge, relative to our own.

It seems that the rulers of the Martian civilization had established a crash program to discover and develop alternative energy sources, and that the directives to the researchers were very simple and direct: identify the form of matter that contains the most energy.

Moreover, due to the absence in the Martian civilization of any form of mathematics higher than basic arithmetic, no purely theoretical approach to this problem was possible. It would have to be done, sample by sample (in much the same way that Thomas Edison tried to find a practical light bulb).

On the plus side, the Martians were possessed of extremely finely nuanced sense organs. So detailed was the data provided by their marvelous sense receptors, the inhabitants of the Red Planet were capable of assessing not only the mass of any given sample of matter, but also the potential energy locked within.

Because of this felicitous twist in evolutionary development, the procedure followed for the energy research program was very basic: collect the sensory observations, divide the amount of potential energy observed by the respective mass, and display the amounts in a table.

Once all feasible materials were examined in this manner, the answer to the Martian energy crisis would leap from the table with all the power of direct perception: the numerically largest ratio would signal the material that would save their dying civilization.

Before even a modest fraction of the total possible samples were examined, however, an interesting observation was made. All of the ratios were identical! Equally startling, and very baffling, the value of each of those ratios was equal to the square of the velocity of electromagnetic radiation in a vacuum!!

In other words, for matter of all types whatever, the amount of *energy* locked in that sample *divided by* the *mass* of the sample was *equal to* the *square of* the *velocity* of *electromagnetic radiation* in a vacuum.

Once it was suspected that this ratio held universally, regardless of the nature of the sample of matter, it was proposed that it be elevated to the status of a law of nature and expressed as an equation between the two basic values which varied and their ratio which was always the same.

But what a cumbersome equation! And the Martians didn't have a lot of energy to spare for such tongue-twisting locutions, so the Law of Conservation of Verbiage was invoked.

It was then proposed that letters be used in place of the words and phrases to which the arithmetically-bound creatures had hitherto been limited. This innovation spread like wildfire and led to the development of the method by which a deductive validation was eventually found for the Law of Mass/Energy Equivalence.

Such was the strange account of the cockeyed development of the theoretical sciences on Mars, as revealed to Einstein by his visitors.

The burning question now being restlessly pondered by a small group of historians on our own planet is this: To what extent was Einstein's own work in physics made possible only by what he learned from the Martians?

Einstein was fond of saying "Chance favors the prepared mind." Was his extensive preparation in need of an extraterrestrial "jump-start," or could he have come up with $E = mc^2$ on his own?

The most reasonable (though by no means decisive) indication that he could have done so lies in the power of what Einstein knew so well from his school days, but which the Martians were able to grasp through only the most excruciating efforts in an emergency situation – namely, algebra.

This is the gist of the growing consensus among the few who are aware of what is known as "The Martian Conundrum." But perhaps this belief is itself merely an artifact of the bias in our own civilization toward deductive, abstract approaches in seeking new knowledge.....

– Gerbis Rosell, Director
Terran Historical Deconstruction

Postscript – light speed and mass-energy equivalence: The most frightening things around today are AIDs and the hole in the ozone layer. Either of them would take years to kill you if you were exposed to them. But I grew up in the shadow of a (potentially) much speedier killer, nuclear holocaust.

As long as I can remember (since the early 1950s), some of the people around me have been scared silly that we might all go up in a big mushroom cloud. Or die shortly thereafter from radiation poisoning. (See Neville Shute's book, *On the Beach*, and the movie based on it.)

Of course, nuclear Armageddon never happened. We're still here. At least, until some terrorists succeed in smuggling a homemade H-bomb into our home town. Or into Disneyland or the Super Bowl (see Tom Clancy's *The Sum of All Fears*). With the collapse of the Soviet regime, nuclear terrorism is probably the only way such a disaster might still happen.

What made these fears possible, of course, was Einstein's great discovery, symbolized as $E = mc^2$. Very simply, this means that the amount of energy in a given quantity of matter is equal to the product of the amount (mass) of that matter and the square of the speed of light.

Find a way to convert all this matter into energy, and a vast power source is yours. And today, of course, we have both weapons and power plants that operate on this principle (those that haven't been decommissioned, that is).

How did we learn of this? Was someone tinkering around with a lump of matter and boom! Lots of energy suddenly appeared? No, indeed. It wasn't at all like the discovery of gunpowder or dynamite or TNT or any of the other great explosives. And for a good reason.

With atomic energy, we are not talking about chemical reactions you can set up in your basement. We are not dealing with observable, perceptual-level entities, but with reactions occurring on the microscopic level, which are thus invisible, unless a very great many of them occur in close proximity to one another in a very short time.

No, our route to nuclear knowledge was very indirect. It involved much theoretical speculation and manipulation of symbols and equations, before the first glimmer of practical applications appeared.

It was not through a series of repeatedly verified laboratory experiments, but by a long process of deduction involving numerous physical concepts and principles, by which Einstein arrived at his deceptively simple result: $E = mc^2$. (For details, consult any standard college physics text.)

The concepts and principles that comprise our knowledge of nuclear physics are quite a ways up the chain of ideas from observation of everyday perceptual reality. Thus, it's not likely that we could ever have arrived at Einstein's equation inductively. Consider all the data to be gathered, and observations and calculations to be made, in order to induce the matter-energy equivalence:

1) Measurements of both the mass (m) and the energy (E) contained in a given quantity of matter for many different kinds of substances.

2) Examination of the ratios and the square roots of the ratios between E and m.

3) Noting the equality (constancy) of these ratios across all the different kinds of matter.

Chapter 4 – How the Martians Discovered Algebra

4) Measurements of the velocity in a vacuum of light as well as all the other different forms of electromagnetic energy.

5) Noting the constancy of those velocities (*c*) across all the different kinds of electromagnetic energy.

6) Noting the equality of *c* and the square roots of the ratios between *m* and *E*: $c = (E/m)^{1/2}$.

Only then (!) could an inductive Einstein have *deduced* that $c^2 = E/m$ – and thus that $E = mc^2$.

But it's rather far-fetched to imagine an inductively talented scientist measuring masses and energy releases in such a systematic way. Certainly the question of their relationship might occur to him. But how would he know how to initiate the energy releases, without all the knowledge leading to, and including, Einstein's equation?

Even with the technology of today to facilitate his work, I don't think an inductively inclined scientist would have had the incentive to amass all this data, unless he already knew (or suspected) what he was looking for.

No matter how the Einstein equation is arrived at, however, it raises interesting questions and implications that I have never seen addressed. For instance, the speed of light in a vacuum is always, at least in theory and observation to date, the same quantity, a constant, *c*, which is approximately 186,000 miles per second. (Sorry…I *loathe* the metric system.)

We might first ask: why is it a constant? What in nature makes the speed of light be what it is, and how? (Note: this is not the same as the question of whether the speed of light is the ultimate physical speed limit in the universe, a staple of science fiction novels.)

And is the speed of light really constant, or is there a faulty premise behind Einstein's equation, such as the rejection of the "ether" (a thin, hard-to-detect physical medium through which electromagnetic radiation is propagated).

Consider also the fact that if *c* is a constant, so is the square root of the ratio, for *any* substance, between energy and mass. Material entities are such that their mass and energy are related in this way.

But what does this mean, ultimately? Is the speed of light determined by the nature of the energy and mass attributes of material entities? Or are those energy and mass attributes determined by the speed of light? Or are both determined by another, more basic fact about material entities? Or are they both correlative, causally irreducible aspects of material entities?

What we *can* say is that material entities have several important attributes that are quantitatively related by Einstein's equation. Each and every physical entity in the universe is such that the square root of the ratio of its energy and its mass is a constant value that is equal in quantity to the velocity, also constant, of electromagnetic radiation emitted by any and all physical entities.

Suppose we think of these aspects of what a physical entity *is* – i.e., its *attributes* of mass, stored energy, and ability to emit or absorb electromagnetic radiation – as some of the basic things it can *do*. We can then see how thoroughly interconnected not only these, but *all* the various aspects of an entity's nature are – i.e., how much of an ontological unity an entity really is.

This meshes well with Ayn Rand's view that Existence is Identity, that a thing *is* its attributes. It also fits nicely with Aristotle's view that cause-and-effect is the Law of Identity applied to action. It is quite a contrast, however, from John Locke's view of an entity as a bundle of attributes, a sort of metaphysical pincushion, into which various attributes are stuck like pins.

When attributes are viewed as discrete, accidentally connected features, rather than essentially connected and inseparable aspects of a thing's identity, the ontological unity and the lawful behavior of entities must remain a complete mystery, a baffling miracle. This is one of the implications that the arch-skeptic Hume drew from Locke's view.

One wonders: how would Hume react to the revelations of modern science? Would he continue to claim that the myriad, comprehensive interconnections of attributes across the entire field of an entity's nature were just contingent and non-necessary? Or would he abandon the atomistic, pin-cushion view of attributes and see instead the indivisible nature of things that exist?

I said earlier that AIDs and the hole in the ozone layer were the scariest things around. Well, although most people are blissfully unaware of it, modern philosophy is actually worse.

Hume, Kant, and their followers have all but wrecked the field for those seeking rational guidance for their lives – and provided the philosophical base for monstrous political regimes such as the Nazis and the Soviets. (See Leonard Peikoff's *The Ominous Parallels*.) This in turn has generated the impetus behind the development and proliferation of nuclear weapons.

Luckily, we have had people like Henry B. Veatch, Mortimer J. Adler, and Ayn Rand to combat these ideas and to offer us sane alternatives. The influence of their clear thinking on the foundations of science and mathematics may yet pave the way for a revitalization of the Scientific Revolution. Similarly, for the libertarian ideals of the American Revolution.

Perhaps someday there will be another Einstein – another person who will lead us to even greater insights into the nature of the universe – someone who will help us to understand the basic nature of matter and energy to an even more fundamental level – someone who will explain why the speed of light (and the mass-energy ratio) is the value it is, rather than some other value.

Postscript 2 – <u>is the speed of light an absolute limit?</u> In reflecting further on this reflection on the speed of light, I'm reminded of stories about the days of air travel before aviators managed to break the *sound* barrier. Back then, people said you couldn't break the sound barrier, because the turbulence around Mach 1.0 would supposedly cause the planes to break apart.

So far as I know, there is no parallel *mechanical* impediment to FTL travel, no physical stresses that would threaten to rip a spacecraft apart as it approached the light barrier. Instead, the inability of space travelers to exceed the speed of light is claimed to be a consequence of Special

Relativity, which is our current comprehensive theory about motion in the universe.

Quoting the entry from Wikipedia:

> Special relativity reveals that c is not just the velocity of a certain phenomenon – light – but rather *a fundamental feature of the way space and time are tied together*. In particular, special relativity states that it is impossible for any material object to accelerate to light speed.

The reason it is impossible to accelerate to light speed, let alone to exceed it, is that as the velocity of a material object approaches the speed of light, its mass approaches infinity. So, according to Wikipedia, *c* is a fundamental feature of *the way matter and energy are tied together*, as I have also argued in the fictional portion of this chapter.

Because of this metaphysical linkage between matter and energy, the amount of energy required to accelerate a spacecraft all the way to the speed of light would thus also be infinite, and that would require as much energy as, or more energy than, the total amount in the universe – depending on whether the universe itself is infinite or finite.

There may (stress: *may*) be a way to skin the FTL cat, but that would entail finding a "shorter distance" between two points than what we currently know to be the shortest route through "regular space." There is no evidence for such "wormholes," however, so this remains an arbitrary idea, baseless speculation, and little more than the product of wishful thinking.

The point is that, without something *like* a wormhole, Special Relativity – which, by the way, is well in keeping with Aristotle's and Ayn Rand's view of motion as being relative, not absolute – dictates that the speed of light is the upper limit of the velocity of anything, matter or energy, travelling through space.

In order to suspend or inactivate (?) or otherwise circumvent the speed of light as a limit, we would have to find a way to create a zone where matter and energy did *not* relate to each other in the same constant way. Whether something like wormholes will turn out to be a feasible way of doing so, or something else we haven't yet imagined would be needed, I don't know.

What I do know is that imagination per se will not get us to the stars in anything faster than a sub-light-speed vehicle, unless that imagination is coupled with intellectual insight into the deeper nature of the cosmos. I eagerly look forward to seeing what the next few generations of theoretical geniuses will come up with.

Chapter 5: Mathematics as an Inductive Science

> *Mathematics is the substance of thought writ large, as the West has been told from Pythagoras to Bertrand Russell; it does provide a unique window into human nature. What the window reveals, however, is not the barren constructs of rationalistic tradition, but man's method of extrapolating from observed data to the total of the universe. What the window of mathematics reveals is not the mechanics of deduction, but of induction.* [Leonard Peikoff, Objectivism: The Philosophy of Ayn Rand, p. 90.]

I have long believed that, in mathematics, induction – the process of arriving at generalizations from more specific observations – is far more important (or, at least, more *interesting*) than deduction. (For an overview of this perspective, refer back to: Introduction, Confessions of a Would-Be Mathematician.)

Not that deduction is insignificant. It is the engine of *proof*, after all. However, without induction, mathematics – like any other discipline – simply could not get off the ground.

In contrast, my favorite philosophy lecturer, Leonard Peikoff, has repeatedly characterized mathematics as being essentially unlike the physical sciences in its basic method. For instance, in his 1983 lectures on *Understanding Objectivism*, he claimed that mathematics is atypical epistemologically – it is deductive, whereas the typical pattern of human knowledge is inductive.

In the 30 years since Dr. Peikoff presented this lecture course, I have not seen any indication that he has modified this view, and he confidently repeated it in a more recent lecture courses on induction. Despite having studied physics in college, however, Peikoff simply does not know a great deal about mathematics, or he would never have made such a claim in the first place.

Perhaps Peikoff is thinking specifically about geometry, but even so, there are mathematically trained Objectivists who have convincingly made the opposite case. Thankfully, in particular, one of Peikoff's associates with the Ayn Rand Institute has emphatically corrected the record.

Pat Corvini presented a course in 2005 called "The Crisis of Principles in Greek Mathematics." The blurb in the Ayn Rand Book Store catalog says that Dr. Corvini "demonstrates that the actual history supports a proper view of mathematics as an inductive science." So, *quod erat demonstrandum* for that little controversy!

To be fair, Peikoff may have been thinking of mathematicians such as G. H. Hardy, the great and rhetorically flamboyant 20th century British mathematician. Here is a perfect example of a mathematician championing "pure mathematics," a theorist who was offended by the idea of practical applications being the yardstick of an idea's value.

Hardy was a giant in the field of number theory, co-writing a classic text in that field. A couple of quotes from *A Mathematician's Apology* clearly portray his bias toward theory: "I am interested in mathematics only as a creative art. Pure mathematics is on the whole distinctly more useful than applied. For what is useful above all is technique, and mathematical technique is taught mainly through pure mathematics."

However, I don't believe that Hardy championed "pure deduction" in mathematics, i.e., starting out with arbitrary postulates and deducing logical consequences. He certainly appreciated rigor and the need to nail down one's intuitive insights and hunches, but his approach was first and foremost inductive and/or intuitive, rather than deductive and logical.

What was "puristic" about Hardy's perspective was not a preference for induction over deduction or vice versa, but the fact that he eschewed practical applications for the realm of "pure" theory. He was scornful of the idea that, to be worthwhile, mathematics had to have a "cash value" in the real world.

In this respect, Hardy was in tune less with the "rational productivity" of Objectivism, and more with the "purely rational" nature of Aristotle's man who, "by his nature, desires to know."

Chapter 5 – Mathematics as an Inductive Science

None of this, however, is meant to denigrate the value or usefulness of deduction to validate mathematical conclusions, and to firmly situate them within a structure of more general and more specific conclusions. The point is merely to underscore the fact that mathematical discovery is primarily *inductive*, not deductive.

Let me give a "simple" example that I think will illustrate this point. Suppose I wanted to demonstrate to you, or help you to discover, that the expansion of $(x + 1)(x - 1)$ is always equal to $x^2 - 1$, no matter what the value of x.

I could just have you apply, *deductively*, the polynomial multiplication rules thusly: $x^2 + 1x - 1x - 1^2$, and you would see that the two middle terms always drop out, leaving $x^2 - 1$. This is a pretty quick way to demonstrate the expansion $(x + 1)(x - 1)$ is $x^2 - 1$, no doubt about it.

But what use would there be in having seen that? Apparently, none, other than having learned a formula for use in simple problems such as: what is the area of a rectangle whose length is 2 units longer than its width? Typically we would multiply w (width) times $(w + 2)$ (length) and get $w^2 + 2w$.

For instance: if width is 9 and length is 11, then the area, expressed in terms of the width is $9(9 + 2)$, which is $9(9) + 9(2) = 81 + 18 = 99$.

This is really the long way around, when we could instead just plug in the numbers 9 and 11 to the w times l formula and get 99 simply and directly. Not too useful, to be sure.

However, let's take a different, more interesting approach. Suppose instead we note that there is another quantity x, in between the length and width, and in relation to which length is $x + 1$ and width is $x - 1$, and that the area of the rectangle is $(x + 1)(x - 1)$ or $x^2 - 1$.

We can generalize: the area of a rectangle with length two units longer than the width is one less than the square of the number that is *one* unit longer than the width – i.e., than the area of the square whose sides are one unit longer than the rectangle's width. For instance, 7(5) is one less than 6(6), 9(7) is one less than 8(8), 21(19) is one less than 20(20).

Now, we can see some usefulness to using expansions in multiplication, even of numbers, rather than just letters. Suppose you want to multiply 99 times 101 (aka find the area of a rectangle 99 units wide and 101 units long). Instead of "simply" multiplying these numbers (in your head, if you're really smart), you can square the integer in between them and subtract 1. For instance: $100^2 - 1 = 10,000 - 1 = 9,999$.

Now, that, I would say, is a *benefit*, a useful application of expansions. It falls in the category of what some call "speed math" (see Chapter 1). And it is just one of numerous examples.

The catch, though, is that you have to have the *insight*, the abstract capacity to "see" that this squaring a number then subtracting 1 is the same as the multiplying the next larger number by the next smaller number – and that the former is *easier*, that it takes less time and effort, than the latter.

My belief, after years of pondering this issue, is that "seeing" such economies or useful applications is much more difficult when you approach a particular idea deductively than when you approach it inductively. That is, it's easier to "see" the payoff, the practical value, of ideas when you grasp them inductively.

I'll try to illustrate this with the same example, the $(x + 1)(x - 1)$ expansion. Suppose we lay out several columns of numbers:

x	$x + 1$	$x - 1$	$(x + 1)(x - 1)$	x^2
1	2	0	0	1
2	3	1	3	4
3	4	2	8	9
4	5	3	15	16
5	6	4	24	25
6	7	5	35	36
7	8	6	48	49

By "inspection," we can see "with the overwhelming clarity of direct perception" that there is an inexorably repeating pattern tying these columns together. The square of a given integer, x, is always going to be 1 greater than the product of the next larger and next smaller integer. In algebraic terms: $x^2 = (x + 1)(x - 1) + 1$, or $(x + 1)(x - 1) = x^2 - 1$. Or, so it appears.

We can quickly verify that this is so by means of the deductive technique shown above. But before we do that, we note that *if* it is a true generalization, there is a very important and useful spin-off: if we want to multiply any two integers that differ by 2, all we need do is square the integer between them and subtract 1.

Many people, for instance, know that $12(12) = 144$. Now they can easily multiply $11(13)$ in their heads: $144 - 1 = 143$. (For other examples, see above or make up your own.)

This is the payoff of induction, I think. It takes at least talent, if not genius, to grasp "speed math" from the deductive route. But with induction, grasping "speed math" takes no particular talent, just curiosity, study, and insight.

More specifically, it requires, first, a puzzle or question to motivate you – then displaying and comparing entries in a table of numbers – then noticing a pattern in the array of numbers ("seeing what you get" when you do thus and such to a particular number) – and finally having the sudden realization that this can be a useful, time-saving technique.

In general, this is how creative discovery occurs in math – and I can personally vouch for the effectiveness of this approach.

I myself have made some rather interesting, if obscure, discoveries in this manner – such as a new way to generate Pythagorean triples (integers of the form $x^2 + y^2 = z^2$). And I suspect that the *meaning and usefulness* of those discoveries would never have occurred to me, even if I had somehow managed to stumble on their mathematical expressions by the deductive route.

Various theorists including George Polya (whose works I recommend highly) see it this way, too. They stress the inductive, creative side of mathematics, especially for young (high school and college) students.

I can't help but think how different the world of science and math would be – and our lives in general – if this was how students were taught algebra, with lots of the roll-up-your-sleeves, immerse-yourself-in-specific-details, and then generalize-from-concrete-instances kind of work.

A monumental example is the incredible Indian genius Srinivasa Ramanujan (1887-1920). Sponsored at Cambridge University by Hardy, he was able to make these kinds of creative discoveries in his head. He filled notebooks with mind-cracklingly abstract equations that numerous professors and graduate students have spent decades trying to verify.

To my knowledge, none of Ramanujan's theorems have been disproved, and a number of them have resisted solution to this day. I can only shake my own in awe at the raw cerebral power that creates and produces such results. For details on his amazing and tragically brief life and career, see Robert Kanigel's biography, *The Man Who Knew Infinity: A Life of the Genius Ramanujan*.

Postscript – Peikoff, Groarke, and Kornblith on induction: Resolving the question of the validity of induction in grounding our knowledge of reality – whether in science or mathematics or philosophy or everyday life – has been one of the long-standing challenges to Ayn Rand's philosophy of Objectivism.

For those concerned with this missing cornerstone to the foundations of human knowledge, I have three very strong recommendations to make.

One is Leonard Peikoff's lecture series "Induction in Physics and Philosophy" (2002). IPP is probably the best overview I have seen of how induction works in building up massive systems of knowledge, whether in science or philosophy, and I think it is some of Peikoff's best work to date.

[Note: these lectures are available, as are those of an earlier course, "Objectivism through Induction," for an extremely nominal fee as digital downloads from The Ayn Rand Bookstore's "e-store." Peikoff's associate, physicist David Harriman recently edited and turned IPP into a book entitled *The Logical Leap*.]

Chapter 5 – Mathematics as an Inductive Science

Peikoff aims to clarify the nature of induction, its relation to our knowledge, and the way in which it operates in science and philosophy. In Lecture 1 of IPP, he characterizes the Problem of Induction thusly:

> How can man know, across the whole scale of time and space, facts which he does not and can never perceive?...When and why is the inference form "some" to "all" legitimate? What is the method of valid induction, the rational method, which alone can prove the generalizations to which it leads? In short: how can man determine which generalizations are true, in other words, they correspond to reality, and which ones are false, they contradict reality? [Leonard Peikoff]

In my opinion, Peikoff meets this challenge head-on and very successfully. In regard to the nature of induction, I think he has hit a bases-loaded home run.

I also enthusiastically recommend Louis Groarke's book *An Aristotelian Account of Induction? Creating Something from Nothing* (McGill-Queens University Press, 2009). His devastating critique of the empiricist-Humean view of induction (Chapter 3, "A 'Deductive' Account of Induction") decisively buries the myth that "inductive inference is not logically valid."

Reading Groarke's book on induction is a very rich and rewarding experience. He covers many issues related to induction, including Cartesian doubt and "brain in a vat" arguments. By far the most intriguing and innovative part of his book, however, contains his argument that induction is actually a special form of deduction.

Groarke calls out Hume and the modern establishment philosophers for their laziness and ignorance about their ancient and medieval predecessors. He shows not only that induction is valid and provides a solid foundation for deductive thought and science, but that it truly *is* "deductively valid."

One of Groarke's extended examples of inductive syllogisms with convertible terms is particularly relevant. It directly faces down the ubiquitous canard that the existence of black swans in Australia proves that inductive inference from one billion white swan sightings to "All swans are white" is not logically valid, and therefore inductive inference is not logically valid.

My third recommendation is a rather short book by Hilary Kornblith entitled *Inductive Inference and Its Natural Ground* (MIT Press, 1993). In Kornblith's words, the focuses of the two main sections of the book are:

1) "What is the world, that we may know it?" "Are there real kinds in nature, or are kinds, instead, merely imposed upon nature by the human mind?" – and

2) "What are we, that we may know the world?" What is the structure of the mind that explains "what it is about human beings which makes it possible for us to have inductive knowledge"?

The blurb on the back cover of the book reads:

> Hilary Kornblith presents an account of inductive inference that addresses both its metaphysical and epistemological aspects. He argues that inductive knowledge is possible by virtue of the fit between our innate psychological capacities and the causal structure of the world. Kornblith begins by developing an account of natural kinds that has its origins in John Locke's work on real and nominal essences. He then examines two features of human psychology that explain how knowledge of natural kinds is attained. First, our concepts are structured in a way that presupposes the existence of natural kinds. Second, our native inferential tendencies tend to provide us with accurate beliefs about the world when applied to environments that are populated by natural kinds.

I like this book a lot, and I urge readers interested in the subject of induction to give it a read. I especially draw your attention to the following:

1) Chapter 3, in which he develops Boyd's idea that real kinds are "homeostatic property clusters."

2) Chapter 4, in which he argues (contra Piaget, Posner, Luria, Quine, and Ayn Rand) that there is *no* evidence available that children *ever* rely entirely on the superficial characteristics of and perceptual similarities between objects in classifying them. – and

3) Chapter 5, in which he shows how errors in inference offer not evidence of its unreliability but insight into how we are able to

arrive at accurate conclusions about the world so quickly and easily (in parallel to how visual illusions are regarded).

Like me, some of my readers have cut their epistemological teeth, so to speak, on Rand's *Introduction to Objectivist Epistemology*. We have also witnessed the Objectivist movement's extended growing pains in coming to grips with the "Problem of Induction."

A key hurdle in solving this problem the clash between the so-called "intrinsicist" Aristotelian view that there are *real* essences in things, upon which we base our concepts and the "objectivist" Randian view that essences, while *based* on real characteristics in things, are epistemological, in the sense of being *contextually identified* and thus subject to revision.

For some time now, I have thought it reasonable to distinguish between real or metaphysical essences (in the Aristotelian sense) and "definitional" or epistemological essences (in the Randian sense). In the latter, Rand incorporates both metaphysical (i.e., causal) and epistemological (i.e., explanatory) factors.

Rand is making the very important point that causally deeper aspects of something are explanatorily more basic as well. But didn't Aristotle see it this way, too, in his work on definitions?

More importantly, isn't science actually getting us to some of these real essences? For instance, isn't the chemical-atomic nature of water its real essence? We don't need to know about quarks or Rand's "little stuff" or Peikoff's "energy puffs" in order to have grasped the real essence of water, do we?

In general, doesn't anti-reductionism (i.e., anti-eliminative reductionism) save us from the supposed dilemma of never being sure we have the "ultimate cause" or "true nature" of something? For example, aren't its chemical-atomic properties the true nature or real essence of water, and won't that always be so?

It may be interesting to know that H_2O is made of 6 up-quarks and 4 sideways quarks, while hydrogen peroxide is made up of 4 up-quarks and 6 charm-quarks. But that would no more reveal the real essence of water than would the ultimate biological units (DNA molecules) the real essence of humans – which for millennia has been seen as possession of a rational faculty!

Postscript 2 – Joseph and Coffey on induction: Some of the Aristotelian philosophers of epistemology and logic from a century ago seem to have had a clearer understanding of the nature of induction than mainstream academic philosophers, or even some Objectivists, do today

I assume that motivated readers have, or have access to, H. W. B. Joseph's *An Introduction to Logic* (1906), so I'll start by recommending that they read (or re-read) pages 177-180 and chapter 18 "Of Induction" and chapter 19 "Of the Presuppositions of Inductive Reasoning: The Law of Causation."

However, one brief passage from pp. 400-401 is well worth sharing here. Joseph begins by quoting J.S. Mill from his *System of Logic*:

> Why is a single instance, in some cases, sufficient for a complete induction, while in others myriads of concurring instances, without a single exception known or presumed, go such a very little way towards establishing an universal proposition? Whoever can answer this question knows more of the philosophy of logic than the wisest of the ancients, and has solved the problem of Induction. [John Stuart Mill]

Joseph then comments: "However we may think of the knowledge possessed by the wisest of the ancients, the question which Mill asks is no doubt an important one. By what right do we ever generalize from our experience? and how can we tell when we have a right to do so?" After discarding syllogism and enumeration, Joseph continues:

> The answer is that all induction assumes the existence of connexions in nature, and that its only object is to determine between what elements these connexions hold. The events of our experience are no doubt particular, but we believe the principles which they exemplify to be universal; our difficulty lies in discovering what principles they exemplify; in that, a close study of particular facts will help us; but were we to be in doubt whether there are any such principles or not, no amount of study of particular facts could resolve our doubt. [H. W. B. Joseph]

This last comment is a nifty example of the kind of insight applied by Rand, Branden, Peikoff and others in analyzing what they call the Stolen Concept Fallacy. The concept of "doubt" only has meaning and validity in relation to principles about which one has no doubt. Without the latter, there would be no need of, and no logical basis for, any such concept as "doubt."

Also, doubt as we may that a particular technique or argument or procedure will give us reliable knowledge in a given case, or that a particular conclusion or hypothesis represents an accurate grasp of the facts of reality, systematic doubt is self-defeating. The best we can do is the best we can do, correcting our mistakes and reaching for deeper insights into the nature of the world.

As for the basis of the "connexions in nature," Joseph says that what underlies the observed same characteristics in things is the same causal conditions obtaining. When the causal conditions vary – e.g., environment or nutrition or mutation causing swan feathers to be black rather than white – the characteristics vary.

It is precisely the *causal* context that you must be very careful to specify, at least implicitly, in your generalizations. If you overlook a relevant part of that context, and you come up with anomalous observations, you must seek to understand the context on a deeper level and revise your generalization accordingly.

None of this means that induction is invalid. Instead, it shows how we must proceed to form valid inductions, and to correct errors that occur.

A less widely known Aristotelian logician and philosopher from the same time period as Joseph is Peter Coffey, professor of logic and metaphysics at Maynooth College in Ireland. His books on metaphysics, epistemology, and logic, written in the early 20th century, are long since out of print, but (very well) used copies may still be found through Alibris and other sources.

I am going to share and comment on a few passages from volume 2 of his logic treatise, *The Science of Logic, An Inquiry into the Principles of*

Accurate Thought and Scientific Method (1912). These are all particularly relevant to the nature and validation of induction.

In the first passage, Coffey draws a sharp distinction between concrete, enumerative universals and abstract, scientific universals. He explains how only the latter can provide necessary principles and laws about the nature of reality:

> We have already repeatedly distinguished between the mere concrete, collective, enumerative universal, and the really scientific universal which is an abstract judgment, embodying some more or less necessary principle or law (95, 195 [vol. 1]). It is this latter that scientific induction proper aims at establishing. (27)

> [Enumerative induction] will not be valid unless the enumeration is complete. The enumeration must be [Greek expression] as Aristotle expresses it; else the argument will be fallacious: there will be an illicit process of the subject of the conclusion. St. Thomas likewise insists that as long as we base our conclusion on enumeration the latter must be complete. So long then as we concentrate our attention on the mere enumeration of instances, and disregard their nature, we can never be certain of our conclusion until we are certain that our enumeration is actually complete.

> Secondly, even where the enumeration of instances is complete, the process does not lead to scientific knowledge, i.e., the knowledge of a strictly universal conclusion embodying what can be called a law. And the reason is manifest. The conclusion expresses a simple addition of instances, and is, therefore, simply a collective proposition whose subject is an actual whole; whereas the strict universal proposition, the abstract universal, can be reached only by generalization of the abstract judgment which establishes some sort of necessary connexion of attributes between subject and predicate. Adding parts to parts, to form a natural whole, gives us a collective idea. Considering an object in the abstract, apart from its individualizing characteristics, putting it into relation with its concrete realizations, actual or possible, indefinite in number, seeing that it is predicable of all, is to universalize and to make scientific progress. For "all science is of the universal and necessary" [Aristotle, Posterior Analytics, i, 5 (5-7)]; i.e., it is expressive of necessary, and

therefore universal, relations between the objects of our thought. The strict universal is no mere actual collection; it is applicable to an indefinite number of instances. Therefore, this kind of induction [enumeration] does not put us in possession of scientific or necessary truth. (29-30) [Peter Coffey]

Get this? Science is universal and necessary, not merely "without exception and by coincidence," as per the Humean skeptics. The latter is the will-o'-the-wisp that is rightly rejected by their critique of "enumerative induction." But that has nothing to do with valid generalizations and science.

The following is my favorite passage from Coffey:

> We find it sometimes stated by modern logicians that the only way of ascending from the particular to the general, explicitly treated by Aristotle, and the only way known to the mediaeval Scholastic logicians, was that of enumerative induction, "complete" and "incomplete"; that we find in these authors no trace of the method of modern scientific induction, the method of attaining to the universal by analysing a limited number of instances and seeking therein a connexion of content, of attributes, a causal connexion, in the nature of the phenomena considered. (33) [Coffey]

Coffey then quotes Joyce's *Logic* to explain why modern philosophers mistakenly thought that Scholastic logicians thought the essence of induction was enumeration, and failed to realize that genuine induction has been around for well over 2000 years:

> The error seems to have arisen from the fact that the most famous of the Scholastics (St. Thomas, Albert the Great, Scotus) do not employ the term induction as the distinctive name of the inference by which we establish universal laws of nature. Following the terminology of Aristotle...they called it proof from experience. The significance of the term induction was somewhat vague. It covered all argument from the particular to the general...It was by a later generation that the term induction was restricted to its present signification. Incautious readers, finding in certain passages the inductive syllogism described as the formula of inductive argument, jumped too hastily to the conclusion that the mediaeval philosophers rested their

> knowledge of the laws of nature on no basis but enumeration. (33, emphasis added) [Coffey]

Coffey continues to drive his point home:

> Now, from the very fact that Aristotle and the Scholastics considered it possible to reach a truth about "all," actual and possible, known and unknown, by an acquaintance with "some," they must have recognized a method of ascent to the "all," other than enumeration. And so they did: viz., the method nowadays known as Physical or Scientific Induction.
>
> When, therefore, we hear it stated that Scientific Induction is an achievement of the modern mind, we must not infer that it was entirely unknown to the ancients. That to modern thought the honour was reserved of seizing upon the full significance of the method, and of applying it with such marked success, even the most ardent defenders of Aristotle and the Scholastics need not deny. But that the principle of this method was known to the latter, their works give unmistakable evidence. [He cites Aristotle, Aquinas, and Scotus.] (33-34) [Coffey]

For all its imperfections, Coffey tells us, there is a real inductive tradition behind modern science that goes back over two thousand years. Thanks to his efforts, and those of others working in the Aristotelian and Thomist traditions – and despite the best attempts of the Humean skeptics and other naysayers to bury it with sophistry and obfuscation – it is alive and well.

From this, it's clear that the modern tendency to saddle Aristotelians with the view of induction as essentially a process of enumeration rests on an *over-generalization*. Moderns have been in a bit much of a hurry to discard the supposed strait-jacket of Aristotelian logic and philosophy of science. Let's slow down and take our time and do it right. We have all the time in the world.

Chapter 6: Equations as Propositions: Using Aristotle to Prove Euclid

> *In his* Logica Ingredientibus, *Peter Abelard argues that the simple affirmative categorical proposition "A human being is white"* (homo est albus) *"should be analysed as claiming that that which is a human being is the same as that which is white* (idem quod est homo esse id quod album est). [Sara L. Uckelman, *Aristotelian Syllogistics with Abelardian Truth Conditions*, p. 1]
>
> > *For Boole, the logical form of a proposition such as "Every square is a polygon"...is really an equation, two terms connected by equality...Here equality is the strictest mathematic equality, "is-one-and-the-same-as," also called numerical identical. Roughly speaking, as incredible as this may seem, Boole would have said that "Every square is a polygon" is better thought of as "Every-square-is [one-and-the-same-as] a-polygon," or better as "All-squares is [one-and-the-same-as] some-polygons... For Boole, the only connector was...the so-called is-of-identity, which is one of the meanings expressed in normal English by the two-letter word "is" as in "One plus two is three" or "Twain is Clemens." The propositions involving the is-of-identity are often expressed by expletive sentences to emphasize the nature of the connector. For example, "Mark Twain is Samuel Clemens" is sometimes expressed by the expanded expletive sentence "Mark Twain is the same person as Samuel Clemens."*
> > [John Corcoran, "Aristotle's *Prior Analytics* and Boole's *Laws of Thought*," History and Philosophy of Logic, 24, pp. 271-272]

As a sort of sneak preview of my forthcoming book on the nature of propositions, I am going to share with you readers some fresh thoughts about how mathematics relates to theory of knowledge – and as is frequently the case, the impetus for my own views derives from the relatively sparse writings on the subject by Objectivist philosophers.

When you want something done, you have to do it yourself...as the saying goes.

In *Introduction to Objectivist Epistemology*, Ayn Rand discussed at some length the parallel or analogy between concept-formation on the one hand and arithmetic and algebra on the other. However, at that time, she expressly declined to comment on how concepts are organized into propositions, and she never took up the task in the remaining 15 years of her life – and to date, no one else in the Objectivist movement has provided a model of propositions that is consistent with her theory of concepts.

[Harry Binswanger and Gregory Salmieri recently published their respective views in *How We Know: Epistemology on an Objectivist Foundation* (TOF Publications, 2014) and "Conceptualization and Justification" in Gotthelf and Lennox, *Concepts and Their Role in Knowledge: Reflections on Objectivist Epistemology* (University of Pittsburgh Press, 2013). I will defer comment, however, to my forthcoming book, *What's in Your File Folder?*]

My working definition of "proposition," for the purpose of this essay, is: *a mental grasp of a fact about the existent(s) referred to by one concept (the "subject-term") by means of asserting or denying the identity of that existent(s) with – i.e., the sameness of that existent(s) and – one or more of the things referred to by another concept (the "predicate-term")*. By "identity" here, I mean not the *nature* of something (what it is), but the *sameness* of that thing with itself

For instance, the proposition "Ayn Rand is the author of *Atlas Shrugged*" is the form in which one mentally grasps *the fact that* Ayn Rand is the author of *Atlas Shrugged* by correctly asserting *the identity of* the thing referred to by "Ayn Rand" with the thing referred to by "author of *Atlas Shrugged*." In other words, the thing referred to by "Ayn Rand" is the same thing as the thing referred to by "author of Atlas Shrugged."

I have only one qualification to make on this definition: any categorical propositions which do not already explicitly exemplify this definition can be restated in equivalent terms to do so. (This is called "translation" into "Standard Propositional Form" or "S is P," for short.)

For instance, "Ayn Rand wrote four novels," translates into: "Ayn Rand was a writer of four novels," which asserts the identity of the thing referred to by "Ayn Rand" with one of the things referred to by "writer of four novels."

When I say that propositions *can* be translated into standard form, I don't mean to say that they always *must* be. For normal conversational purposes, there is no need to do so whatever. However, for various reasons in dealing with logical issues, it is highly advisable to do so. And in general, if there's any doubt, they *should* be converted to standard form.

Now, the stumbling block in virtually all writings on proposition theory has been the modest little device of the *copula*, the verb "to be" which links subject and predicate. In what sense *is* the subject of a given proposition "the same as" or "identical to" the predicate?

Consider these examples: Socrates is a man, men are animals, my car is red, etc. Is there any common element in these and other uses of the "is" of *identity*? Also, what about the claim that the verb "is" or "exists" functions primarily to imply *existence*, and if so, what about propositions about things that do not exist?

Equations as propositions

Setting aside these and numerous other questions for the present, I thought it would be helpful to take a closer look at one of the more naïve attempts to explain propositions by likening them to mathematical equations.

For example, in saying "The Morning Star is the Evening Star," the claim goes, we are really more simply saying, "The Morning Star = [equals] the Evening Star," in the same sense that saying "5 is 8 minus 3" is more simply said as "5 = 8 minus 3."

The problem is, it is *not* the same sense. Not precisely. Propositions are not a kind of equation. However, equations *are* a kind of *proposition*! Specifically, they are propositions about *numbers of units*.

A proposition, when you regard it in most basic terms, is an *identity* claim – or a denial of such identity, in the case of negative propositions. It is the assertion (or denial) that the thing or things referred to by the subject term of the proposition is/are identical to (i.e., *is/are the same thing in reality as*) the thing or things referred to by the predicate term of the proposition.

In other words, a proposition is the claim (or denial) that if you examine the one or more things designated by the subject term, you will find a *match* for that thing(s) among the one or more things designated by the predicate term. This matching process is the fundamental cognitive process that underlies grasping and forming propositions.

For instance, "My car is a Ford Focus." If you look at the complete group of things referred to by "Ford Focus," you will find that the thing referred to by "my car" is the same thing as one of them. My car is a match for one Ford Focus. (But not conversely, of course.)

"Ford Focuses are automobiles." All of the things referred to by "Ford Focus" are the same things as some of the things referred to by "automobile." Each Ford Focus is a match for one automobile.

"Ayn Rand is Alissa Rosenbaum." The person referred to by "Ayn Rand" is the same person as the person referred to by "Alissa Rosenbaum." Ayn Rand is a match for Alissa Rosenbaum.

Which takes us to the frequently offered example: "The Morning Star is the Evening Star." The thing referred to by "The Morning Star" is the same thing as the thing referred to by "The Evening Star." Once we learn that they are both the planet Venus, we learn that they are a match, that they are the same thing, and that a proposition which asserts this identity is *true*.

As stated by Ayn Rand, the things referred to by a concept are *units*, i.e., things regarded as being members of a group that have one or more similar characteristics. Seeing things in this unit-perspective is the basis for our being able to form concepts.

The terms used in subjects and predicates are sometimes concepts of a plurality of things, such as apples, human beings, horses, Ford Focuses, planets, etc., but sometimes they refer to specific individual things, such as my Ford Focus, Ayn Rand, etc. Whether we say that these latter terms stand for units or merely "items," however, the process of comparing and matching subject and predicate is the same.

Do the units or items referred to by the subject term match those referred to by the predicate term? If so, then the proposition is true; if not, then it is false. (And if the proposition is *denying* that subject and predicate are identical, then it is the *absence* of a match that makes the proposition true, and the presence of a match that makes it *false*.)

This perspective works for attributes, as well. We might say "my true love's hair is black (colored hair)." But more poetically, we say the same thing with attributes in subject and predicate: "Black is the color of my true love's hair."

If you look at all the instances of the color referred to as "black," you will find that one of them is identical to the color referred to as "the color of my true love's hair." One instance of black color is the same thing in reality as the color of my true love's hair. They are a match.

This also works for numbers and quantities. For instance, when we say: "There are three pencils on my desk," we are actually intending to say: "The number of pencils on my desk is (an instance of the number) three."

Another example: if my weight were 200 pounds (I wish), I could express it with quantities in subject and predicate: "(The quantity of) my weight is 200 pounds." One instance of the quantity "200 pounds" is the same thing in reality as the quantity of my weight. Or, I could express it with numbers in subject and predicate: "200 is my weight in pounds." One instance of the number "200" is the number of my weight in pounds.

This brings us to the issue of numbers and equations. While Rand chose not to go into depth about propositions in her book on concept theory, she did give a brief comparison of them to mathematical equations (1990, 75):

> Since concepts, in the field of cognition, perform a function similar to that of numbers in the field of mathematics, the function of a proposition is similar to that of an equation: it applies conceptual abstractions to a specific problem.

This is exactly correct and for a very good reason: equations *are* propositions, in which the left side of the equation is the subject and the right side is the predicate.

An equation asserts that there is a match between two numbers, one or both of which is the result of a mathematical operation. We can state this assertion of a match between numbers in a mathematical proposition (equation) as the identity of subject (the left side of the equation) and predicate (the right side).

For instance, if we have a collection of units (or whatever), the number of which is 5, we can compare that number to the number of units in the collection resulting from the removal of 3 units from a collection containing 8 units. We see that the number of units referred to by "5 units" is the same number of units as the number of units referred to by "8 units less 3 units."

In other words, when we see that we have a match in this instance, we can express it in abbreviated form as "5 is 8 minus 3," or "$5 = 8 - 3$." Here, both the "is" and the "=" mean: "is the same thing in reality as," or: "is the same number of units as."

Naturally, when we want to manipulate numbers of units, stating the processes in verbal, propositional form would be *far* too cumbersome: "the number of units referred to by '5 units' is the same number of units as the number of units referred to by '8 units less 3 units." It is much more compact and easy to grasp and deal with when that identity is stated "$5 = 8 - 3$."

However, the crucial point to grasp here is that equations *are* propositions, propositions about numbers of units, and that is their fundamental tie to epistemology and proposition theory in particular. (Inequalities are *negative* propositions about numbers of units.)

To reverse this process and to claim that propositions are a kind of equation (or inequality) is as nonsensical as claiming that concepts are a kind of number. In the former (propositions), we are gathering and thinking about units, things regarded as similars, while in the latter (equations), we are gathering and thinking about *numbers* of units.

To put it another way, in the former, we are concerned with a one-to-one correspondence (or lack of same) between units, while in the latter, we are concerned with a one-to-one correspondence (or lack of same) between *numbers* of units.

Now, this might sound like numbers of units and units are different enough that they are not that closely related after all. But what they have in common is that the equations and propositions using them are actually pointing, respectively, to *instances* of units and *instances* of numbers of units.

"My car is a Focus." The thing (entity) referred to as "my car" is the same thing as *one instance* of the things (entities) referred to as (the automobile) "Focus."

"Black is the color of my true love's hair." The thing (attribute) referred to as "the color of my true love's hair" is the same thing as *one instance* of the things (attributes) referred to as (the color) "black."

"$5 = 8 - 3$." The thing (number) referred to as "$8 - 3$" is the same thing as *one instance* of the things (numbers) referred to as (the number) "5."

Applying this model of the proposition may seem rather simple as I have presented it here. And truly it is (as I will show in my forthcoming book).

It applies not only to the full range of positive, negative, universal, particular, and singular propositions, but also to axioms (i.e., axiomatic propositions) that state basic facts of reality, such as "Existence exists," and to propositions about things that do not exist, such as sea serpents, the present King of France, etc.

I invite readers to test this claim for themselves. If they do, then they may wonder, as I have for some time now, precisely what the advantages are of the brain-cracking constructs of modern logic, and how generations of academics could have justified the salaries earned by making human knowledge more difficult to acquire and to understand than it needs to be.

Categoricals vs. conditionals in logical arguments

Now, here's a further application, in mathematics, of this model of the proposition. Suppose you're trying to lay out a mathematical proof, like the ones we did in high school geometry. There are two basic ways of stating the steps in such arguments, but there is disagreement about which kind is better to use.

The traditional way is to put them in what's called categorical statements, in the form "x is y" or "all x is y" – for instance, "all squares are rectangles" or "all triangles are plane figures with 3 sides."

The more modern way is to use conditional statements, in the form "if x, then y" – for instance, "if something is a square, then it is a rectangle," "or if something is a triangle, then it is a plane figure with 3 sides."

At best, the modern logicians say, using categorical statements and arguments is at best unnecessary and should be avoided, because such propositions and syllogisms are logy, long-winded, verbose, "constipated," old-fashioned, and just plain more difficult than those that use conditionals.

This claim seems a bit odd, when you consider the relative simplicity and economy of the categorical and conditional premises in these two syllogisms:

All cows are mammals   vs.   If x is a cow, then x is a mammal.

Bossie is a cow   vs.   Let x = Bossie, such that Bossie is a cow.

Therefore, Bossie is a mammal   vs.   Therefore, Bossie is a mammal.

At worst, the moderns claim, categoricals are often not even sufficient. For instance, they say, you cannot use them to prove propositions from Euclidean geometry, let alone items from higher math like the integral calculus, Gödel's Incompleteness Theorem, or the Heine-Borel Compactness Theorem. (Don't ask; I don't know this Heine-Borel thing either.)

Well, here is how I handle this challenge, in regard to proving Euclid's Book I, Proposition 2, which is: to place a straight line equal to a given

Chapter 6 – Equations as Propositions

straight line with one end at a given point. (Take out your pencil, paper, straight-edge, and compass, if you want to follow along.)

First, though, let me say that when I took high school geometry, I don't recall using any logical processes other than straightforward deduction from axioms, definitions, and previously established results. What few conditional statements the proofs may have contained could easily be rephrased as categoricals, and that is how I will proceed here.

Given and having constructed the following:

1) Given segment BC and outside point A.
2) Connect A and B, making segment AB.
3) Make circle 1 on B with radius AB.
4) Make circle 2 on A with radius AB.
5) Make circle 3 on B with radius BC.
6) Extend chord DB from intersection D of circles 1 and 2, through B to point G on circle 3.
7) Make circle 4 on D with radius DG.
8) Extend chord AD from D to A to point L on circle 4.

Now, the assignment is to prove that the construction has been successful and, therefore, that segment AL = segment BC. Here is the proof:

1) BC = BG (all radii of the same circle are equal)
2) DL = DG (all radii of the same circle are equal)
3) DA = DB (Proposition 1)
4) AL = BG (from 2 and 3: all equal segments diminished by the same amount are equal)
5) AL = BC (from 1 and 4: all segments equal to the same segment are equal

As you can see, despite Euclid's requirement of constructing 4 circles and 2 (or 3) line segments, the process for proving his proposition is relatively simple, requiring only two deductions. So, I did not have to stand on my head or get out a "buggy whip" in order to coerce my brain into laying out the proof.

Also, the deductions are both categorical syllogisms, using categorical propositions, and the premises (in parenthesis) are all categorical as well. (Radii of the same circle are equal. Equal segments diminished by the same amount are equal. Segments equal to the same segment are equal.) Basic Aristotelian deductive logic applies nicely to Euclidean geometry.

And now for the qualifications and objections...

First of all, Aristotle would be quick to admit that more is needed in order to establish Euclid's propositions than deductive logic and categorical propositions and syllogisms. One must discover or intuit the deductive pathway, perhaps by "thinking backward" from the intended conclusion, to figure out what is needed in order to establish the conclusion.

Nowadays, we call this "reverse engineering." How do I get this result? What if I did this? What result would I get? Once we find the right logical pathway, we can cast the result in categorical terms, rather than hypothetical or conditional. But this is not difficult, and it is not new, and it is not controversial.

Secondly, though, where are the categorical syllogisms? The five-step proof given above was highly compressed, with some of the premises in parenthesis, some of the conclusions also serving as premises, etc. In total, there were five syllogisms, all in the form of "Barbara" (or bArbArA, where all three propositions are A, or universal positive, propositions):

>BC and BG are radii of the same circle.
>
>Radii of the same circle are equal.
>
>1) So, BC and BG are equal.
>DL and DG are radii of the same circle.
>
>Radii of the same circle are equal.
>
>2) So, DL and DG are equal.
>DA and DB are sides of equilaterial triangle (ADB).
>
>Sides of an equilaterial triangle are equal. (Proposition 1.)

3) So, DA and DB are equal.
Equal segments diminished by the same amount are equal.

AL and BG are equal segments (DL and BG) diminished by equal amounts (DA and DB). (2 and 3)

4) So, AL and BG are equal.
Segments equal to the same segment are equal.

BC and AL are equal to the same segment (BG). (1 and 4)

5) So, AL and BC are equal.

Naturally, once we automatize the various definitions and propositions and the ways in which they are deductively related, we don't need repeat all these steps explicitly. Nor should we do so, except for pedagogical purposes, such as training young math students to "show their work."

But once the method of proof has been mastered, mental shortcuts and notational economies and abbreviations are normally and reasonably permitted, and this does not somehow make the arguments non-categorical or otherwise fishy. If we hadn't been granted such boons, high school geometry would surely have taken four years instead of just one!

Another pedagogical purpose, of course, is showing good-faith skeptics that one has "dotted the i" and provided an airtight categorical proof of, for instance, Euclid's proposition I-2. For instance, in a single step of the above proof, say #5, we refer to steps 1 and 4 and draw on the principle that segments equal to the same segment are equal.

Now, just because this is abbreviated does not mean that you are not using deductive logic, with categorical syllogisms. So, making the implicit explicit can be illuminating, especially for those who do not see the connections in full detail or who need to ingrain them in their thinking. But it is work, and if you want to see an old man sweat, well, you came to the right place!

And no, we're not done yet. Now we have to address the question of whether the propositions themselves were in categorical form: All S are P, No S are P, Some S are P, Some S are not P. These are the officially authorized forms for categorical syllogisms.

Take, for instance, the conclusion of the first syllogism in the above proof: "BC and BG are equal." Now, what does this mean? Is it like saying: "Plato and Aristotle are men"? This is a condensed way of saying: "Plato is a man," and: "Aristotle is a man." If so, then it is saying: "BC is equal," and "BG is equal." But this is nonsense.

Surely it is instead saying something more like: "Plato and Aristotle are friends," which more clearly expressed is: "Plato and Aristotle are men who are friends to one another." And indeed, when we restate the conclusion of the first syllogism as: "BC and BG are line segments that are equal in length to one another," we see that this is precisely what it means.

What we learn from this is that it is vitally important to say exactly what we mean – or at least to *know* exactly what we mean, and to be able to spell it out more explicitly, if doubt arises.

However, there should be no controversy about the ability of Aristotelian logic to deal with binary relations like "are friends of one another" or "are line segments equal in length to one another," and there should be no serious objection to abbreviating these expressions as "are friends" or "are equal."

A further quibble involves the fact that, for instance, line segments BC and BG are individuals (individual line segments), not categories – and the statement about them is like saying: "Bill and Mary are equal," rather than: "Men and women are equal." (Yet, nitpickers will pick nits!)

This is true enough, but it does not rule out such propositions for use in categorical syllogisms. If it did, then the Socrates syllogism would be out the window.

However, traditional practice for centuries has been to reword "Socrates is mortal" as: "All Socrates is mortal," treating Socrates as belonging to a single-member class, and there is good justification for it, justification that goes deeper epistemologically than the logical formalisms of "All S are P," etc.

Chapter 6 – Equations as Propositions

What we are saying in "Socrates is mortal," or more precisely: "Socrates is a mortal being," is that *what is referred to by "Socrates" is the same thing in reality as one of the things referred to by "a mortal being."*

This is exactly the same cognitive process with exactly the same epistemic justification as in saying "All men are mortal (beings)," which is to say that *the things referred to by "All men" are the same things in reality as some of the things referred to by "mortal beings."*

This approach fits in very well with Ayn Rand's unit-perspective model, viewing a single-member group as being like a "mental file folder" (Rand's analogy) containing one item along with a lot of data about it, similar to a "dossier." The same is true for two-member groups, such as the one containing line segments BC and BG in the Euclid I-2 proof.

(Whether or not we want to label these single- or double-member groups—such as "Bill" or "Bill and Mary" or "line segment BC" or "line segments BC and BG" as "concepts," they function in the same way as concepts when used in a categorical proposition.)

So, expressing the first syllogism in this way, and spelling out the relationship more clearly, will be a bit less pretty than simply saying: "BC and BG are equal," but here it is: "All BC and BG [of which there are two in the present construction] are line segments that are equal in length to one another." And fleshing out the entire syllogism leading to that conclusion looks like this:

> All line segments BC and BG [of which there are two in the present construction] are radii of the same circle.
>
> All radii of the same circle are line segments of the same length.
>
> So, All line segments BC and BG [of which there are two in the present construction] are line segments that are equal in length to one another.

This is a straightforward "Barbara" syllogism, as are the other four I used above in my categoricals-only proof of Euclid I-2. All categoricals, all the time.

This also shows that the deceptively simple expression "BC and BG are equal" is really just a disguised categorical proposition that is fully capable of incorporating the binary relationship of "equality." We just have to realize that the relationship of equality has been embedded in the category expressed in the predicate.

Consider: "John and Mary are siblings (of one another)." All this requires is that we form the category "people who are siblings of one another."

Since John and Mary are members of this category, we can say: "John and Mary are two [members of the group of] people who are siblings of one another" – or more concisely (understanding it to be just an abbreviation): "John and Mary are siblings (of one another)."

Thus, "siblings (of one another)" is a legitimate predicate for expressing a binary relationship in a categorical proposition. (Although I'm not an expert on ancient Greek or Latin grammar, it seems incredible that Aristotle or the Scholastics would have rejected this as controversial, or as a tortured, *ad hoc* modification of their own perspective and methodology.)

The same would be true for using the category "line segments that are equal in length to one another" – or simply, "equal (line segments)" – as a predicate in a categorical proposition in a Euclidean proof.

It's ironic that some Objectivists look askance at what seem to them to be *ad hoc* categories. They're comfortable with concepts like "people" and "siblings," but not "people who are siblings of one another" – or concepts like "south" and "Chicago," but not "cities that are south of Chicago."

But as Rand said, we properly form concepts (and categories) not willy-nilly, but when the situation requires it. We aren't cognitively required to conceptualize and categorize *ad infinitum*, but only *ad necessitatum*. (I apologize if the latter is not a real word. I just threw it out there, because it seemed helpful.)

As for transitivity of equality and symmetry, my use of them is not "begging the question." My task was to start with everything given and proved up to that point and to then prove Book I, Proposition 2. When I undertook to do this, I did so on the understanding that I was not required to also prove every assumption that Euclid stated at the outset and used in his proof.

What I *did* do was take any proposition that Euclid or someone else might be more comfortable phrasing as an "if...then" or "since..." and rephrased it as a categorical. "If two quantities are both equal to the same third quantity, then they are equal to one another" becomes: "All pairs of quantities that are equal to the same third quantity are equal to one another."

This amount of verbiage may seem like an overly puristic strait-jacketing of traditional logical procedure. But it is unfortunately necessary to meet all of the sophistic challenges to the power and appropriateness of categorical propositions and syllogisms. This includes the claim that categorical propositions cannot legitimately replace conditional propositions in arguments.

Now, we have just cleared a big hurdle in showing the efficacy of categorical propositions and syllogisms in mathematical proof. The gutsy thing to do might thus be to plunge on ahead and show how similar efforts can establish the integral calculus, Gödel's Incompleteness Theorem, etc. But that is not our responsibility.

We have done enough. We have shown that it is in principle possible. If it's really all that desirable for our critics to see everything non-hypothetical expressed as a categorical, before they will concede that categoricals are not somehow inferior to hypotheticals, then it is well past time for them to stop denigrating categoricals, to roll up their sleeves, and to help with the heavy lifting. (As opposed to simply standing by, idly watching an old man sweat!)

Because Aristotelian syllogistic logic and categorical propositions suffice for *all* deductive arguments that are not purely hypothetical in nature, they also suffice for mathematical arguments. As has been explained earlier in this chapter, this is because, more fundamentally, all mathematical equations can be recast as categorical "S is P" propositions, from "$2 + 2 = 4$," on up to the most complex equation you can imagine.

Again, just to demonstrate the point: the equals sign in mathematics simply means "is," or "is the same as." "2 + 2 = 4" means that the number of things referred to by the combination of two things and two more things is the same (thing in reality) as the number of things referred to by the numeral "4."

It's another way of saying that "2 + 2" and "4" are equal, and we just saw that the binary relationship of equality can readily be embedded in a predicate. "The number of things referred to by the combination of two things and two more things and the number of things referred to by '4' are equal (to one another)."

Mathematicians (and math students!) are not to be blamed if they prefer the vastly simpler condensation of "2 + 2 = 4," but they would do well to remember that the simpler expression is not just an empty formalism, but a way of expressing the relationship of numbers to reality.

So, if a mathematical idea is expressed (or re-expressed) as an "S is P" proposition, then its proof can be expressed in terms of "S is P" categorical propositions and validated with categorical syllogisms. No "If S then P" propositions and conditional syllogisms are needed!

And that applies as well to the second sentence preceding, which can be recast as a universal positive categorical: "All mathematical ideas that are expressed as 'S is P' propositions are mathematical ideas whose proof can be expressed in terms of 'S is P' propositions and validated with categorical syllogisms."

True, there might not be any practical advantage in doing so in a given case. For instance, I used exactly the same number of words in converting the above conditional to its corresponding categorical.

The point, however, is that mathematics is, in principle, ontologically groundable in this way, and *only* in this way. Any conditional that cannot be validly converted to a categorical is not a claim to knowledge of reality, but only speculation.

If, as the moderns seem to prefer, this ontological grounding function of propositions is rejected, then mathematics – not to mention intellectual activity in general – is, at best, a sometimes inadvertently useful activity of establishing logical connections between symbols that are little more than a congeries of floating abstractions. And Hume, the arch-skeptic, has won.

Postscript – when not to use conditionals: Conditional propositions and conditional syllogisms are the language of exploration, of *looking for connections*. Actually, I would be surprised if mathematicians did *not* use conditionals at least *some* of the time – specifically, during their thought processes and speculations.

Conditionals are the "training wheels" of conceptual thought, the scaffolding of trial-and-error, the mental lubricant of experimentation. Once the connections are found and validated, however – whether in mathematics or science or philosophy or everyday experience – it is our *option* to leave them in conditional form or to re-cast them in categorical form.

For instance, once we scientifically reach (thanks to Newton) the insight that: "If a body is unacted upon by an outside force, its state of motion will continue unchanged," we are entitled to replace it with: "All bodies unacted upon by an outside force will continue unchanged in their state of motion."

As for syllogisms, leaving them in conditional form, even when there is really no unfulfilled premise (like "if it rains tomorrow"), is useful for showing the points where the speculative process took place during the construction of the proof. That can certainly be helpful sometimes for pedagogical purposes or for reviewing one's thinking during a long chain of reasoning.

For most practical purposes, however, conditional talk sounds like unnecessary hedging. It sounds as if one is unable or unwilling to trust the process of induction for some reason.

For example, it really sounds odd to offer up as a premise: "If x is a horse, then x is a mammal," rather than: "All horses are mammals," or more simply: "Horses are mammals." When I hear the former locution, I want to ask: "You mean you don't already know that horses are mammals? Did you skip induction class that day?"

Similarly, it sounds odd to say: "If x is a circle, then its radii are equal in length," rather than: "The radii of a circle are equal in length." (The latter itself is a more economical way of saying: "The radii of a circle are line segments that are equal in length to one another.")

This generalization states a fact, and once you know that fact – or any other fact – you *know* it. You don't have to keep nervously blurting out "if…then," clinging to your conditional woobie, when you state it.

Admittedly, Euclid does appear to use conditional talk here and there. He says things like: "Since AB is equal to FB and BD is equal to BC, triangle ABD must be congruent to triangle FBC." But this is equivalent to saying: "AB is equal to FB and BD is equal to BC. Therefore triangle ABD must be congruent to triangle FBC." If "since" is conditional, then so is "therefore."

Either way of speaking is just a way of making explicit the causal relationship between premises and conclusion, i.e., the fact that the premises are the reason for (the *conditions* of) the conclusion. "Because (by reason of the fact that) AB is equal to FB and BD is equal to BC, triangle ABD must be congruent to triangle FBC."

Just because this is the fully spelled out meaning of a categorical syllogism does not somehow make it non-categorical. It just means that when it is appropriate, we can strip away all the extraneous conditional or causal verbiage and lay out the bare structure of the argument, understanding that "x and y, therefore z."

I personally don't care how mathematicians find it most felicitous to express the steps of their reasoning, but when the rubber meets the road, their conclusions are in the form "S is P," not "If S, then P." And if they are sure enough about their conclusions to express them as categoricals, what's wrong with doing the same with their premises?

(Again, we're not talking about hypothetical arguments such as: "If he went to the party, he got drunk. He went to the party. Therefore he got drunk.")

I really do understand why modern logicians prefer to say: "If x is a man, then x is mortal," rather than: "All men are mortal." It's the old Existential Import bugaboo. Perish forbid we should "imply" that something exists when it doesn't really exist, as in: "All sea monsters are reptiles." Much better to play it safe and say: "If x is a sea monster, then x is a reptile."

There is a very simple reason why inferences using propositions about non-existent things have a tendency to go awry. It's because the *mode of existence* or "universe of discourse" predicated of the subject term is not spelled out.

The corrective is to state the full, explicit meaning of the proposition: "All sea monsters are real creatures that are reptiles." Yes, this is claiming (not implying) that sea monsters really exist. But all we have to do is simply (1) examine the units or items in the subject and predicate categories, (2) observe that there is no match between them, and (3) declare the proposition false and move on – not wring our hands over its (allegedly) committing a fallacy.

Modern logicians have had the option of manning up and doing this for all universal propositions, but instead they have wimped out with their "if x, then y" evasions. They were like the teacher who punishes the whole class for the difficulties caused by one student, rather than figuring out how to appropriately deal with that student.

It would have been so easy for them to have finessed such outliers as sea monsters and "the present King of France" with minor, common sense modifications of traditional logic. Instead, the moderns tore apart Aristotle's Square of Opposition and immediate inference and left us with hypothetical universal propositions and a premature obituary for traditional logic.

This is progress? I don't think so.

Chapter 7: Much Ado about Nothing: Zero as an "Operation-Stopper"

The point about zero is that we do not need to use it in the operations of daily life. No one goes out to buy zero fish.
[Alfred North Whitehead,
An Introduction to Mathematics, p. 63]

Arithmetic, not to mention the more abstract levels of mathematics, is utterly dependent on reality, and it cannot be stressed enough that its validity is tied to its traceable derivation from reality. If that relationship were not traceable, then any application it had to the real world would be purely accidental.

Modern mathematics has, to a great extent, replaced the human race's common-sense cognitive connection to reality with floating abstractions and arbitrary constructs. This has given rise to numerous mathematical and logical paradoxes.

In order to reverse the long detour taken by modern mathematics, and to properly orient mathematics (back) to the real world, a lot of re-interpretation is necessary. Some of that re-interpretation has to do with getting clear about exactly what operations we are or are not performing.

This essay will explore some of the mathematical functions of the number "zero," with special emphasis on its role as an *operation-stopper*. Part of my fascination with this issue is due to my intense desire to get completely clear about what is going on whenever "zero" is used in mathematics.

(I am also driven in part by how irritated I am by the standard interpretation of the "empty set," as if it is an actual container with some special kind of thing called "nothing" in it. This idea is behind modern logic's attempt to put mathematics on a sound logical basis, which unfortunately involves the illogical idea of combining something with nothing in various ways. See Chapter 8.)

So, in this chapter, we will seek to clarify how the use of zero as one of the inputs of an arithmetic or mathematical expression often used to denote a mathematical operation actually *prevents* any such operation from taking place. We will see that this is so, even when and even though there is conventionally a *nominal* expression of a result, *as if* some such operation had taken place.

Very simply, as I say, I have this notion that – despite appearances and despite conventional wisdom to the contrary – zero is an "operation-stopper."

I came up with the idea over 17 years ago, while taking David Kelley's cyber seminar on propositions. Some of the seminar participants wondered what zero powers refer to in reality, and what is the ontological meaning of "x^0."

I remarked at the time that the zero power does not refer to a number of things in reality, but to a number of times an operation is supposed to be performed on the unit 1. (Being the multiplicative identity, 1 is an implicit factor in *any* arithmetic expression).

Some of the seminar participants agreed with me that my explanation of what is going on with zero *exponents* (see Chapter 3) was very clarifying, but it wasn't until several years ago that I became curious to see how far I could extend the idea of zero as an operation-stopper.

Starting, of course, with basic arithmetic, it seems to apply straightforwardly to counting, addition, and subtraction. For instance:

(1) If someone asks me to count the number of blue chairs in my living room, I don't tell them "zero." I say there aren't *any* chairs there.

(2) If I leave the room empty of blue chairs, I don't say I have "added zero blue chairs."

(3) If there were three blue chairs and I left them in the room rather than removing them, I don't say I have "subtracted zero blue chairs."

In the above examples, there has been no counting and no addition and no subtraction. "Zero" is a "red light" to those operations.

This idea of zero as an operation-blocker is not a blunt instrument, however. You do have to be careful to specify what operation has been blocked, and in what way, and this varies significantly from one kind of operation to another.

Chapter 7 – Zero as an "Operation-Stopper"

Now, yes, it is true that if I have three blue chairs, and I remove one, and then add two, and then remove one, the *net effect* certainly *appears* to be the same as if no blue chairs were added or removed – or, in the conventional way of speaking, that "zero" chairs were added or removed.

But note carefully, the appearance of "zero" chairs having been added or removed is simply an expression of the fact that two opposing kinds of very visible operations on existing chairs have *canceled each other out* – not that some invisible operation was performed on non-existent chairs!

It's very clear (because I was there!) that actual blue chairs *were* removed, then added, then removed. Three operations were performed that canceled each other out.

The net effect is *equivalent* to chairs not being added or removed at all – except for the fact that I know from direct experience that it is *only* equivalent and not *actually identical* to chairs *not being* added or removed. (And in any case, it is neither equivalent *nor* identical to "zero" chairs *being* added or removed.)

Another example: You can't be the parent of zero children, and you can't be a husband, straight or otherwise, with zero wives! Nor can you *add* zero wives to your life – or zero children to your household.

I want to be absolutely clear about this. If you don't get married or have children, then you *are not adding* zero wives or children to your life or household. If you decide to remain single and childless, you do not say: "I have decided to *add zero* wives and children to my life." You say: "I have decided *not to add any* wives or children to my life."

Similarly, if we consider the conventional discussions of "0" as the "identity element" of addition, we see that no addition is actually being performed.

For instance, when we write out the equation $1 + 0 = 1$, does that represent the *addition* of 0 to 1? That is the standard interpretation. But how can we add nothing to something? Actually, what we are doing is *not* adding *anything* to something.

In other words, the notation *really* symbolizes that we *are not* adding anything to 1, not that we *are* adding 0 to it. The zero means that the operation of adding *is not performed*. We simply count 1, and that's it.

The case is the same for the equation $0 + 1 = 1$. Contrary to the conventional view, this cannot represent the addition of something to nothing. There is *not anything* that we are adding *something* to.

In other words, the notation *really* symbolizes that we *are not* adding 1 to anything, not that we *are* adding 1 to 0. The zero means that the operation of adding *is not performed*. Again, we simply count 1, and that's it.

The objection may be made that 0 is not nothing, because it functions as the identity element of an additive commutative group. But it does so precisely because when 0 is "added to" some other number, the other number is left unchanged.

So, "adding 0" is not *doing something* to the other number. In fact, the "+ 0" is *not doing anything, except affirming that you are not doing anything* to the other number. Of course affirming that you are not doing anything to a number is affirming *something*, but it is *not* affirming that you are *adding*.

0 is the "additive identity" precisely *because* what you start with is identical to what you end up with. When trying to add 0, you end up with what you started with, because 0 is not anything. If it were something, it would have had an effect on what you started with, and you would have ended up with something other than that.

By trying to add 0, in other words, you have not done anything to the original number quantitatively. All that you have done is to *mentally affirm* that you have not done anything to it quantitatively. That is the real meaning of $x + 0 = x$.

So, I do *not* deny that $x + 0 = x$. I just deny that you are *adding*. There is no operation being performed – just an acknowledgement that *when you do not do anything* (adding zero is not doing anything) *to a number, the number remains unchanged.*

That is why 0 is the "additive identity." It does not change the numerical identity of the number to which you *try* to add it, because you *cannot* add it. There isn't anything *to* add!

As a purported addition, the equation $x + 0 = x$ means that the *sum* of x and 0 is x. But the equation is *not* an addition. There is no sum of x and 0, because you are *not adding*, so the *sum* of x and 0 is undefined.

Chapter 7 – Zero as an "Operation-Stopper"

However, the *expression* $x + 0$ *does* equal x. Not because x *plus 0* is x, but because x is x. When you include $+ 0$ in a statement, it is as if you had not said anything after the number preceding it.

(By the way, the same reasoning, of course, applies to $x * 1 = x$. You are not multiplying. You are just stating that x is x. 1 is the *multiplicative* identity, just as 0 is the additive identity. There is no operation being performed, just a recognition that *when you do not do anything to a number* – multiplying a number by one is not doing anything to that number – *the number remains unchanged*.)

So, clearly, I am not advocating that we dispense with 0 as the identity element of addition, any more than I would advocate that we dispense with 1 as the identity element of multiplication. I am proposing instead a *reinterpretation* of what addition to, or with, 0 really means (and similarly, a reinterpretation of what multiplication of, or by, 1 really means).

$0 + x$ does not mean that something is *added to nothing*, and $x + 0$ does not mean that *nothing is added to* something. Instead, they mean that *no* addition is performed, and that we simply have a *count* of whatever is *not* added to 0, or of whatever 0 is *not* added to.

In both $5 + 0$ and $0 + 5$, the result of 5 leaves the (identity of the) original 5 unchanged, and this is precisely because no operation has been performed. It retains its identity as a *count* of 5, because there isn't anything that has been added to it, and there isn't anything that it has been added to.

These seem like simple, sensible re-interpretations of standard practice, if we are to keep mathematics and ontology aligned with one another, and mathematics and our actual mental operations aligned with one another. It certainly seems preferable to treating mathematics like an enormous spider-web of intellectual relations bearing no necessary connection to the real world.

That is what reality-oriented philosophy "contributes to mathematics," not (as some fear) a bizarre, invalid jettisoning of the additive identity element.

Something similar can be seen for "multiplication by zero." Typically, we are taught that any number multiplied by 0 is 0. This is another misinterpretation of what is going on. In $5 * 3 = 15$, we are multiplying 5 by 3, by which we mean we are counting the total number of things in 3 groups of 5 things per group.

However, in the expression 5 * 0, we are not *multiplying* 5 by *some specific number;* instead, we are *not multiplying* 5 by *anything*. Thus, we are not counting anything. In particular, we are not *counting* the total number of things *in some specific number of groups* of 5 things per group; indeed, we are *not* counting the total number of things *in any groups* of 5 things per group.

The notation specifies that there *zero* multiples of 5, i.e., there aren't any groups of 5 things per group. The zero "stops" any attempt to multiply anything by zero.

The *result* of this would-be multiplication is, in fact, zero. But why? Why isn't it instead the number 5, as it would be for 5 + 0? Because the result of 5 * 0 is that *there isn't anything* that we have – there aren't any groups of 5 things per group. Zero is not the *product*, but instead is just a *result* – the result of *observing that there aren't any groups* with 5 things in them.

Considering that multiplication is just compressed addition, you can see this easily: 5 * 3 is 5 + 5 + 5, 3 multiples of 5. The number 5 must appear 3 times as the only addends, and the sum of those three multiples of 5 is 15.

However, 5 * 0 is *not any* multiples of 5. The number 5 must not appear any times, and there aren't any other addends, which means that there isn't any addition (and hence there isn't any multiplication) being performed.

So, please note carefully: 0 is expressed not as the *product*, but as the *result* of 5 * 0. It is not the *product* of an actual multiplication operation, because it is instead *what there is to be counted when no such operation can be performed*. There isn't *anything* to be counted – and specifically, there aren't any groups of 5 to be counted – and this fact is expressed as "0."

Now, what about "multiplication *of* zero"? Again, we are conventionally taught that 0 multiplied by any number is 0 – and again, this is a misinterpretation of what is happening. In 3 * 5, we are multiplying 3 by 5, by which we mean we are counting the total number of things in 5 groups of 3 things per group.

However, in the expression "0 * 5," we are not *multiplying some specific number* by 5. Instead, we are *not multiplying anything* by 5.

Thus, we are not counting anything here. In particular, we are not counting the total number of things in 5 groups which have some specific number of things in each of those groups. Indeed, we are not counting the total number of things in 5 groups with *any specific number* of things per group.

The notation specifies that there are 5 multiples of 0, but this means that there are 5 groups of things that do not contain any things. A group that does not contain any items is impossible. In parallel with the previous case, the zero "stops" any attempt to multiply zero *by* anything.

Again, the result of this would-be multiplication is zero, and again it's because *there isn't anything* that we have – in this case, there aren't (because there *can't* be) 5 groups of things which do not contain any things. Again, zero is not the *product*, but instead is just a *result* – the result of *observing that there aren't* 5 groups *which do not contain any things*.

Once more, to clarify: zero is not the description of some special kind of *thing that is present* in the result of attempted multiplication by zero, but instead the description of the *absence of any thing* that is present in that result.

And again, viewing multiplication as compressed addition, we can see why this must be so: $3 * 5 = 3 + 3 + 3 + 3 + 3$, 5 multiples of 3. The number 3 must appear 5 times as the only addends, and the sum of those five multiples of 3 is 15.

However, $0 * 5$ is *not* 5 multiples of *anything*. There isn't any number that must appear 5 times, and there aren't any other addends, which means that there isn't any addition (and hence there isn't any multiplication) being performed.

Again, note carefully: 0 is expressed as the *result* of $0 * 5$, but this is not the *product* of an actual multiplication operation, but instead is *what there is to be counted when no such operation can be performed*.

Now, is there any legitimate way to establish that 0 is the *product* of *actually* multiplying anything by 0 – rather than, as I claim, merely the *result of attempting* to multiply anything by 0?

It is sometimes claimed, in fact, that $a * 0 = 0$, as a consequence of the distributive law and the definition of subtraction, to wit: $a * 0 = a * (b - b) = ab - ab = 0$. However, this supposed proof actually employs a false premise, in substituting $b - b$ for 0. Thus, despite its flawless logic and the truth of its conclusion, it is only true by accident. Garbage in, garbage out.

Now, before examining this alleged proof, we must first acknowledge that the reverse, *substituting 0 for (b − b)*, would *not* commit a logical fallacy. Such a substitution simply depends on the premise: "all $b − b$ are 0" – in other words, that all results of opposite numbers canceling each other out are absences of any quantity.

This is just as true as the fact that all results of the numbers of chairs brought into and removed from an empty room canceling each other out are absences of any chairs in a room. (In conditional form: "if x is $b − b$, then x is 0" – i.e., if x is the result of opposite numbers canceling each other out, then x is an absence of any quantity.)

But this is not what the traditional argument does. Instead, it does just the opposite, which is illegitimate. Instead of substituting 0 for $(b − b)$, *it substitutes $(b − b)$ for 0*. This substitution depends upon the premise: "all 0 are $b − b$" – in other words, that all absences of any quantity are the result of opposite numbers canceling each other out.

This is simply not true, any more than all absences of any chairs in an empty room are the result of the numbers of chairs brought in being canceled out by the number of chairs removed. (This fact is apparent even if the premise is put in conditional form: "if x is 0, then x is $b − b$" – i.e., if x is an absence of any quantity, then x is the result of opposite numbers canceling each other out.)

Here is a simple analogy for seeing the flawed premise in the traditional argument. First, here is a sound argument that parallels the legitimate substitution of 0 for (b − b):

> Fido is a dog. (true)
>
> All dogs are mammals. (true)
>
> Therefore, Fido is a mammal. (true)

The conclusion of this syllogism substitutes "mammal" for "dog," which is a legitimate substitution, for everything which is true of dogs is also true of mammals (and specifically, all dogs are mammals). Similarly, everything which is true of results of opposite numbers canceling each other out is also true of absences of any quantity.

However, this is not the direction of substitution and logical implication used by the traditional argument for $a * 0 = 0$, which instead uses the illegitimate substitution of $(b - b)$ for 0. Thus, the standard argument for the validity of multiplication by 0 is made according to the same pattern as this unsound argument:

> Fido is a mammal. (true)
>
> All mammals are dogs. (false)
>
> Therefore, Fido is a dog. (true)

The logic of this syllogism is flawless, and the conclusion is true. (Just as the logic is valid and the conclusion true in the traditional argument for $a * 0 = 0$.) However, the second premise is false, for not everything which is true of mammals is also true of dogs, and thus not all mammals are dogs. So, *the argument – while valid and producing a true conclusion – is not sound*. Garbage in, garbage out!

Similarly, not everything which is true of absences of any quantity is also true of results of opposite numbers canceling each other out. Thus, not all absences of any quantity are also results of opposite numbers canceling each other out.

Valid logic that establishes a true conclusion can be seductive, but *valid logic and a true conclusion are no guarantee that the argument is sound*.

For instance, we know that $a * 0 = 0$, just as we know that Fido is a dog. But just as we don't know the latter *because* of the false premise that all mammals are dogs, neither do we know the former *because* of the false premise that all absences of quantity are also results of opposite numbers canceling each other out.

In other words, *the traditional argument for $a * 0 = 0$ is a logical failure*.

An argument that *works* for establishing this conclusion is the one given above, based on my perspective that zero is an operation-blocker: $a * 0$ means that there aren't any groups that contain a units each, and any situation in which this is true is also a situation in which there aren't any units.

Unlike multiplication by a non-zero number, you are not counting groups with units when you attempt to multiply by zero. There are *not any* multiples of the number a when it is "multiplied" by zero.

In other words, you have *not done anything* quantitative to the number *a*. All that you have done is to *mentally affirm* that you have *not counted* any groups containing *a* units, and as a result you *do not have any units*. That is the real meaning of $a * 0 = 0$.

A similar thing happens in regard to the "zero power," which is always 1 for any real number. E.g., 5 to the zero power is 1, 100 to the zero power is 1, etc. Some people are mystified by this, wondering what it means ontologically. Well, its meaning is in the *operation* that *is not* being performed. (In that respect, a zero power functions similarly to a zero addend, as above.)

As I discussed in Chapter 3, the key to grasping what is going on with powers is to realize that the factor 1 is always the base to which the power multiplication is applied or not. For instance:

1) 5^2 *actually* means: the number one multiplied by the number 5 two times.

2) 5^3 means: 1 multiplied by the number 5 three times.

3) 5^0 means: 1 *not* multiplied by the number 5 *any* times.

The zero exponent means: the operation of power multiplication on the factor 1 *is not performed*. That is why any number to the zero power is always 1. Not because that number is *taken* times itself *zero* times, but because 1 is *not taken* times that number *any* times.

So, for a really quick way to grasp the ontology of 0 and how that impacts mathematics, I strongly suggest that we consistently look at *all* powers, including 0, as being based on the unit 1 taken times a certain factor a certain number of times – and that when the certain number of times is 0, you are basically left with the unit 1, to which you have not done anything!

In contrast to my unit-1 perspective, please note that the conventional view's attempted proof that $a^0 = 1$ is just unsound as the one used for $a * 0 = 0$. The supposed conventional proof goes like this:

$$a^0 = a^{n+(-n)} = a^n * a^{-n} = a^n/a^n = 1.$$

As above, the illicit premise, the revelation of which disables this supposed proof, is the erroneous notion that $0 = (n + (-n))$, and that $(n + (-n))$ can thus be substituted for 0. This idea presumes that all absences of any quantity are the result of opposite numbers canceling each other out. This is simply not true, any more than all absences of any chairs in a room are the result of the numbers of chairs brought in and removed canceling each other out.

(This fact is apparent even if the premise is put in conditional form: "if x is 0, then x is $n - n$" – i.e., if x is an absence of any quantity, then x is the result of opposite numbers canceling each other out.)

Contrast this with the traditional approach to explaining the zeroth power. People stand on their heads, manipulating the Associative and Commutative "laws," in a vain (because unsound) attempt to show how the zero power is the nth power minus the nth power.

The bottom line: if they are referring to something other than a *real* mental operation that is *not carried out any times*, they are *not* truly understanding the zero power, as the symbol of the real mathematical operation of factor multiplication *not being carried out* in relation to the unit 1.

Now, a more general point: I am *not* suggesting that we dispense with zero in calculations entirely.

For one thing, zero is an indispensable place-holder in the Arabic system of notating numbers above 9. As we are taught in elementary arithmetic, any number with two or more digits is interpreted as the sum of the count of so many single units and the count of so many groups of 10 units and the count of so many groups of 100 units, etc.

The function of zero as a place-holder thus arises when there is not a corresponding unit or group of units for a given digit in a number. The lack of such a unit or group is represented (place-held) by 0, which signifies that there is not a count of those units or groups. (E.g., 201 = 1 unit + *no* groups of 10 units + 2 groups of 100 units. I illustrate this point further in Postscript 2 below.)

For another thing, zero is an indispensable indicator of the result of subtraction or, equivalently, addition of opposing kinds of quantities (i.e., positive and negative numbers) that cancel each other out, such as in financial accounting, temperature, etc. (I owe this insight to Pat Corvini's lecture series "Two Three Four and All That," and I discuss this point further in Postscript 3 below.)

One prime example is transferring the right side of an equation to the left side by subtracting it from both sides. Since the two sides are set as being equal, subtracting the right side from the left side results in the left side being cancelled out, which we indicate notationally as 0.

E.g., $x^2 + 2x = 3$ becomes $x^2 + 2x - 3 = 0$. The 0 is not some special number, but instead just the symbolic indicator that the $x^2 + 2x$ and the 3 are equal quantities of opposite magnitude and thus cancel each other out when combined.

This is actually just a special case of subtracting a number from itself, which more basically is adding a number and its negative counterpart. E.g., $8 - 8 = 8 + (8 - 8) = 0$. When these two numbers, which are equal quantities of opposite magnitude, cancel each other out, we indicate the result notationally as 0.

This, however, this does not mean that we're left with some special thing called "nothing" or "zero." "Zero" as the result of an addition or subtraction process means *there isn't anything left*.

Or, consider an extended example. The series $17 - 3 - 8 + 4 - 9 - 1 + 5$ can be rewritten so that the subtractions are replaced by additions of negative numbers as: $17 + (-3) + (-8) + 4 + (-9) + (-1) + 5$. After five additions, the first six positive and negative numbers have canceled each other out, and you are left without anything to which 5 is added.

Notationally, this is expressed as $0 + 5$, but this means not that 5 is *added to nothing*, but that 5 is *not added to anything*, for there is not anything to add it to. Therefore, 5 *is not being added* to 0. You simply *count* 5, as if you had begun with 5 in the first place, and that's it.

So, zero has two functions in this example.

1) First, it indicates that a canceling out of positive and negative numbers has occurred, and that nothing is left after that process.

2) Second, it functions as a blocker of the next proposed operation. By indicating that there isn't anything to add the subsequent number 5 to, what happens next is not the *addition* of 5 to 0, but the *beginning anew with a count* of 5, as if that were what we had started with in the first place.

Chapter 7 – Zero as an "Operation-Stopper"

That is an example of how, in general, we should interpret the result x in the identity equations: $0 + x = x$, and $x + 0 = x$. First of all, the result of addition with or to 0 is undefined; the result of such attempted addition is not a sum. The result x is simply a *count* of a number which is not added to anything, or which does not have anything added to it.

Again, *there is no sum*, when 0 is one (or both) of the numbers presented for addition. The zero stops the addition process, and instead there is just a process of *counting* the other number.

You still say that the result of the *attempted* addition process is x (in each case), but that is because the number x was *not added to* anything, or *not added to by* anything. The number following the = sign is not the result of addition involving 0, but represents the non-existent effect upon the non-zero number of not having performed any addition.

(The interpretation of the result x in the identity equations $1 * x = x$ and $x * 1 = x$ is slightly different, but in principle the same as in the case of the identity equations for 0: the result of such attempted multiplication, whether of or by 1, is not a product. The result x is again simply a *count* – in this case, of the number of things in *one* group which has x members, or of the number of things in x groups each of which has *one* member. In each case, what is going on is not multiplication, but simply counting.)

Let's briefly consider some more extended examples. Suppose $x = 14 - 14 + 12$, or $14 + (-14) + 12$. The positive and negative 14's cancel each other out, leaving a result of 0. The 0 then stops any further addition, and the 12 is simply counted.

The result of the string of operations is thus *not* a result of a process of addition, but the result of a process of *counting what there is*, which is only 12.

Suppose $x = 12 + 14 - 14$, or $12 + 14 + (-14)$. We can proceed from left to right and arrive at the result of 12, but that is not controversial. Instead, if we use the associative law of addition, we can first combine the second and third numbers, arriving at $12 + 0$. Again, the 0 stops any further operation of addition, and the 12 is simply counted.

And again, the result of the string of operations is *not* a result of a process of addition, but the result of a process of *counting what there is*, which is only 12.

But even though we say the *sums* of $0 + x$ and $x + 0$ are undefined, we still say that the *result* of an attempt to add 0 and x, or x and 0, is x. The addition operation is blocked, so there is no sum.

All that is there is *a number which has not been added to anything, or which has not had anything added to it*. That is counted, and that count is the result of its not being added to anything, or not having anything added to it.

Suppose we extend the series like this: $17 + (-3) + (-8) + 4 + (-9) + (-1) + 5 + 0$. Per my interpretation, the calculations would end after the first six additions, the last one where I "added" 5 to the 0 net result of the first five additions – i.e., where I didn't have anything to add 5 to, so I just started anew with a count of 5.

Now, although the 5 *restarts* the possibility for addition, the 0 following the 5 *stops it again*, because there isn't anything to add to 5, and the final result of the series remains 5.

Suppose we further extend the series like this: $17 + (-3) + (-8) + 4 + (-9) + (-1) + 5 + 0 + 6$. As above, the result of the adding and counting operations through the final 0 is 5. So, once again, the 5 restarts the possibility for addition, and adding 6 produces a final result for the series of 11.

In this case, as in the preceding, we see that interpreting 0 as an additional blocker produces the same arithmetic outcome as if we thought of 0 as being added to, or being added to by, another number – with the additional benefit that we are no longer thinking in terms of the ontologically contradictory notion of adding nothing to something or something to nothing.

Instead, we are *not adding* anything to something, or *not adding* something to anything. We are blocked from performing an illogical operation. We are merely counting what is there.

Admittedly, the way we are trained in school, when we see or arrive at a zero in a string of calculations like these, is to simply *ignore* the zero and deal with what's left. In our mental operations, we simply pass by the 0 and *do not do anything* with it. In fact, however, what we are automatically thinking is something like: *don't perform any operations with this*.

In addition, subtraction, multiplication, or powers, the something we are not doing to or with the zero differs in specifics, but the general principle is the same. We just have to get clear on what it is that we are not doing anything to, whether in addition, multiplication, or powers.

As an operation-blocker in addition, 0 says: "Don't add anything to the preceding," or: "Don't add what follows to anything," or both. (I discuss what I have figured out so far about subtraction in Postscript 3 to this chapter.)

As an operation-blocker in multiplication, 0 says: "Don't count things in groups, when there are no groups," or: "Don't count things in groups, when there are no things," or both.

As an operation-blocker in exponents, 0 says: "Don't multiply the unit 1 by any factors of the base number."

These are *not* "bizarre metaphysical interpretations," as some might say. They are the result of *directly observing* what we *in fact* do mentally, operationally, when we use zero in addition, subtraction, multiplication, or powers.

You can do something that has a *result* of zero – namely, count – but you cannot do something to zero with something else, and you cannot do something to something else with zero.

In contrast, one truly bizarre idea is the conventional notion regarding multiplication that zero can *do something* to another number that wipes it out, or that another number can *do something* to zero and wipe itself out.

The zero result in attempting to multiply by, or with, zero is not what you "get" from the operation, but what you "have" from the outset, because no operation is possible.

Another truly bizarre idea is the conventional notion regarding addition that zero can *do something* to another number and yet leave it unchanged, or that another number can *do something* to zero and yet leave itself unchanged. The non-bizarre reality of the matter is that if you have not changed something, you have not done anything to it or with it!

(Note: *not done anything* – not: "done nothing," as though nothing were some special kind of "something" you can *do* that has no effect in addition, rather than nothing simply being *not anything, period*.)

This is true not only in mathematics, but also in the real world. Absence of (perceivable) evidence is not evidence of absence! The demons of micro-determinism will surely seek their revenge on anyone who insists on saying, for instance, that an interaction of "barely" touching a brick wall with a rubber glove leaves *either* the rubber glove *or* the brick wall unchanged.

As Arthur Koestler wrote long ago in *The Act of Creation*, modern physics has demonstrated and proved that entities are not the static, unchanging things of traditional, quasi-Platonic physics, Instead, physical objects are seething cauldrons of imperceptible activity, with atoms frequently being exchanged with other entities and the surrounding environment.

(And you thought that when you nudged that other car in the parking lot, you didn't "steal" some of his car's paint!)

There is *no* interaction without physical consequence, however small. Where there is action, there is change. Where there is interaction, there is *some* change in each of the interacting entities. Where there is no change whatsoever, there has been no action. And when there is no change in interacting entities, there has been no interaction.

This is simple metaphysical fact, and the alternative to this is truly bizarre. An action with no consequence is impossible, and so is an operation with no effect.

This is yet another example of how we ignore the connection of mathematics to the real world at our peril. In mathematics, as in actions and processes in the real world, a corollary of the Law of Cause and Effect is: if there is no cause, there will be no effect; and if there is no effect, there was no cause.

Now, the case of division is interesting, in that conventionally, division *of* 0 by any number is permitted and is defined as 0, while division of any number *by* 0 is not permitted and is regarded as undefined. Let's examine these in turn, to see how my perspective on 0 as an "operation-stopper" pertains to division.

First, consider that when we divide a non-zero number by another non-zero number, this is equivalent to multiplying the first by the reciprocal of the second. E.g., $8/(½)$ – i.e., 8 divided by ½ – is the same as $8 * 2$ (i.e, 8 times 2), which means that we are counting the total number of things in 2 groups of 8 things per group, which is 16.

Chapter 7 – Zero as an "Operation-Stopper"

Similarly, $0/(½)$ – i.e., 0 divided by ½ – is the same as $0 * 2$ (i.e., 0 times 2), which means that we are trying to count the total number of things in 2 groups that do not contain any things. As pointed out earlier, this is a contradiction. In reality, we are not counting anything.

Specifically, we are *not* counting the total number of things in 2 groups that do not contain any specific number of things, for there are no such groups to be counted. The zero stops its being multiplied by 2, so we might expect that it would also stop its being *divided by* ½. But does it?

With $8/(½)$, we are basically asking: how many groups of a half-unit each are there in 8 units? We see that for each of the 8 units, we can count 2 groups of a half-unit each, and doing this 8 times, we see that there are 16 groups of a half-unit per group – which is no great surprise, since it mirrors the process of multiplying 8 by 2. (How many units are there in 2 groups of 8 units per group?)

However, with $0/(½)$, we are asking: how many groups of a half-unit each are there in 0 units? We see from the numerator 0 that there are *no* units that contain groups of a half-unit each, so there aren't any groups of a half-unit each to be counted.

Note that the zero stops itself from being divided by 2, just as it stops itself from being multiplied by 2. (How many units are there in 2 groups which do not have any units? There aren't any units to count, so the *resulting count* – not the *product* – is 0.)

The *result* of this attempted operation of division of 0 is 0 – not because that is what we *get from* the operation (viz., its "quotient"), but because it is what we *have from* the outset, because no operation is performed. (There aren't any units to perform the operation on!)

Now, what about division *by* zero?

Let's follow the pattern noted above for division by ½. With $8/(½)$, we are asking: how many groups of a half-unit each are there in 8 units? In each of the 8 units, we can see that there are 2 groups of a half-unit each, and doing this 8 times, we count a total of 16 groups of a half-unit each.

However, for division by 0, something very different happens. With $8/0$, we *seem* to be asking: how many zero-size fractions of a unit can we count in a group of 8 units? This is the standard interpretation, and the standard answer is: uncountably many.

However, as we have seen, 0 does not mean that there are zero-size fractions of a unit in a group, any more than it means there is a special "zero quantity" of chairs in a room. Instead, it means that there aren't any zero-size fractions of a unit in a group, just as it means there aren't any non-chairs in the room. (This problem also arises in considering the notion of an "empty set." See Chapter 8.)

So, what we are *really* asking is: how many *non*-zero-sized fractions of a unit, are *not* there in 8 units? Well, let's see. Wouldn't we just starting counting and then...not...ever...stop? Wouldn't that number of units be uncountably large? Yes, the number of non-zero-sized fractions of a unit that are *not* in 8 is not just very large; it is impossible to determine how large it is.

The *result* of the attempt to divide by zero is conventionally referred to as "undefined," since it is a number that is so large that it is, in principle, impossible to count – which is understandable, once we understand that 0 stops the operation of division and does not allow it to generate a quotient. However, it is incorrect to call the result of (attempted) division by 0 "infinity." Infinity is not the actual quotient of a process of division.

The most we can say is that as the divisor becomes smaller and smaller and thus "approaches" 0, the quotient becomes larger and larger and "approaches" infinity. But when the divisor *is* zero, it blocks the division operation and thus the generation of a quotient.

Furthermore, we know that this is true, no matter how small or large the number that zero tries to divide is. Regardless of the size of the number being divided by 0, we never get a definite result. Instead, no matter what that number is, we always have an uncountably large number of non-zero-sized fractions of a unit *not* contained in that number.

"Infinity" is not what we "get" when dividing by 0, but what we *do not have in* zero, in the form of the endless, uncountable, and thus "undefined" supply of non-zero-sized fractions of a unit that are *not* in zero. "Infinity" isn't how many *non*-units *are* present, but how many *units* are *not* present!

Finally, it may be helpful to see what happens when we try to divide by "infinity" – that is, by an *undefined number* such as 1/0 or 8/0 or 10,000/0. Again, conventionally, *any* number divided by "infinity" is *supposed* to be 0. The reason given is in terms of *limits*. As the divisor becomes larger and larger, the quotient becomes smaller and smaller, and as the size of the

Chapter 7 – Zero as an "Operation-Stopper"

divisor approaches being uncountable, the quotient approaches 0 "as a limit."

This all seems very reasonable: any number (whether 1, 8, or 10,000) divided by "infinity" is 0. Yet, the reciprocal operation, 0 times "infinity" does not yield any number in particular (whether 1, 8, or 10,000). So, now the question is: why does the operation of division by "infinity" seem to work, but the multiplication *of* "infinity" by 0 does not?

I suggest that my approach provides an easy explanation for this anomaly. For instance, what if we try to divide 8 by "infinity" in the form of the undefined number 1/0? Will the answer be 0, and if so, why?

Inverting the attempted division of 8 by 1/0 converts it into attempted multiplication of 8 by 0/1, which we know how to handle. We have seen that since 0/1 is 0 divided by 1, such division is blocked, leaving the result of 0 (there aren't any 1's to count in 0 groups). This means that 8 multiplied by 0/1 is 8 multiplied by 0. Again, 0 is an operation-blocker, this time blocking the multiplication operation, again leaving the result of 0 (there aren't any groups of 8 units per group, and thus no units to count).

This explains why the *result* – i.e., not the quotient, but the resulting count of units – of any operation of division by "infinity" (i.e., an undefined number) is always 0. And it does so not in terms of any dubious application of the concept of "limit," but in terms of the absolute fact that *there are no units to count, so the operations of division and multiplication do not occur!*

Or, suppose we bite the bullet and, instead of using the "work-around" of inverting the attempted division, we look squarely at what is going on when we attempt to divide 8 by an uncountably large (and thus undefinable) number, and give it a straightforward explanation from the perspective of blocking the operation of division per se.

For instance, when happens when we try to divide 8 by 1/0? Remember, first of all, that 1/0 is an uncountably large number. Remember also that when dividing 8 by ½, we are asking how many groups of half a unit each are there in 8 – that when dividing 8 by 2, we are asking how many groups of two units each – etc.

So, when attempting to divide 8 by an uncountably large number, we are asking how many groups of each with an uncountably large number of units there are in 8. *There aren't any* such groups in 8, or in any other number, which means that attempted division of any number by "infinity" (i.e., an uncountably large number) is always 0.

And in reflection, this seems quite reasonable. If "infinity" is an uncountably large number, it is certainly true that 0 is an uncountably *small* number. As fractions get smaller and smaller, they "approach" and get *relatively* closer to being uncountable, but 0 is *absolutely* uncountable. There isn't anything there to count!

So, if attempting to divide by an (absolutely) uncountably small number, 0, will result in an uncountably large (and thus undefined) number, "infinity," then it stands to reason that dividing by an uncountably large (and thus undefined) number, "infinity," results in an uncountably small result, 0.

In summary: these various reframings of the basic arithmetic operations may seem a bit labored and peculiar. I am convinced, however, that they are how our mental processes *really* work in adding, subtracting, multiplying, dividing, and doing powers. I hope this provides sufficient reason for readers to consider it a not only plausible, but actually superior way of viewing these operations.

The reader, in doing so, will hopefully also have seen noticed the shortcomings of the standard, traditional conception of zero. In particular, I hope the reader has become more aware of the ways in which the standard view conception leads to paradox and obfuscates rather than clarifies how our minds do arithmetic (including exponentials).

Unfortunately, however, this is the way some people prefer to do math. They prefer to set up an arbitrarily defined set of rules ("axioms") for manipulating symbols and then play with them, occasionally exclaiming in great surprise when their arbitrarily based manipulations produce a pattern that applies to the real world.

It is just not as easy or fun to acknowledge that mathematics is an *abstraction from* the real world, and that, to be valid, every rule and procedure must be based on or ultimately derivable from some concrete mental operation or other that is directed toward real objects and their attributes, actions, and relations.

(The German mathematician Gödel was a very formidable manipulators of symbols. In an essay I plan to publish in my forthcoming book on logic, I discuss the equivocation in his infamous "slingshot" argument that all facts are the same, single fact. Gödel's reliance on Russell's tottery notion of "definite descriptions" is about as reassuring as libertarians and Objectivists relying on Greenspan's "understanding" of the free market, the money system, and capitalism. Spare me!)

Advances in mathematics have often been accompanied by a lot of protest. Ultimately, however, they have been adopted not necessarily because they were properly understood conceptually, but because they *worked*.

We have a lot of wonderful things in our lives because mathematicians, scientists, and technicians found things that worked. This is so, even while they have not always properly grasping the foundation in reality for the discoveries they made and applied.

As long as they don't blow us up, that's OK. In principle, however, I don't appreciate the knee-jerk brandishing of the Holy Cross of conventional wisdom when people like me try to connect our abstraction ideas to their base in the real world.

Postscript – is zero-to-the-zeroth power undefined? It is conventionally argued that 0^0 is *undefined*, because (it is alleged) for non-zero n, $0^n = 0$, and $n^0 = 1$, but when n is zero, the first equation becomes $0^0 = 0$, while the second equation becomes $0^0 = 1$, and since 0^0 can't be both zero and one, it must be undefined.

It is also conventionally argued (somewhat inconsistently with the preceding) that $a^0 = a^{n+(-n)} = a^n * a^{-n} = a^n/a^n = 1$, provided a is not 0. If a = 0, then we get $0^0/0^0$, which is undefined. (As already mentioned above, I regard this argument as dubious, but see the discussion here: http://www.physicsforums.com/showthread.php?t=39038#post283411)

Because of the long history of controversy over this issue, it would be reasonable to proceed cautiously in endorsing any view. (See the articles on "Exponentiation" and "Empty product" on Wikipedia.)

However, as I have already shown, there are a number of problems with the conventional view of the zero power in general. In view of this, I suggest that, for reasons consistent with my rationale for the nature of the zero power, a strong case can be made for rejecting both of these conventional arguments that 0^0 is undefined and accepting instead the view that $0^0 = 1$.

First, the positive case: if (as I propose) $a^n = 1 * ($...n factors of a...$)$, then $a^0 = 1 * ($...0 factors of a). That is, to say that you multiply 1 by 0 factors of a, is to say that you do not multiply 1 by any factors of a.

That is the real meaning of $a^0 = 1$. You have *not done anything* quantitative to either the unit 1 or the base a. All that you have done to it is to *mentally affirm* that you have *not* multiplied the unit 1 by any factors of a.

Now, suppose $a = 0$. This means that $a^0 = 0^0$, which is interpreted on my model as: *you do not multiply 1 by any factors of 0*. That is why $0^0 = 1$. In this case, you have not multiplied the unit 1 by any factors of 0. As above, you are left with the unit 1.

What about the claim that $0^n = 0$, and thus $0^0 = 0$? Well, again invoking my model, 0^n means: 1 multiplied by n factors of 0, but 0^0: 1 multiplied by 0 factors of 0, which means: 1 *not* multiplied by *any* factors of 0. Again, you have not multiplied the unit 1 by any factors of 0, and once more you are left with the unit 1.

These two results show that, on my model at least, 0^0 is defined and is 1.

What about the claim that when $a = 0$, $a^0 = a^n/a^n = 0^0/0^0 = 0/0$, which is undefined? Setting aside the fact that this argument has a flawed premise (see above), there is another error in this claim.

When $a = 0$, $0^0 = 0^n/0^n$, which means: (1 multiplied by n factors of 0)/(*1 multiplied by n* factors of 0). And when $n = 0$, 00 = 00/00, which means: (1 multiplied by 0 factors of 0)/(1 multiplied by 0 factors of 0). This means: (1 *not* multiplied by *any* factors of 0)/(1 *not* multiplied by *any* factors of 0) – in other words: not 0/0, which is indeed undefined, but 1/1, which is 1.

A possible objection: why should $0^0 = 1$, when $0^1 = 0^2 = 0^3 = ... = 0^n = 0$, for all n not equal to zero? Well, because it does! Why should 0 behave any differently than any other base number taken to the zero power? And why should 0 behave exactly as the base number 1 in all cases?

$0^n = 0$ for all non-zero n, not because you actually *can* multiply the unit 1 by some number of factors of 0, but because *there aren't any groups* that contain 1 unit, and any situation in which this is true is also a situation in which *there aren't any units*.

In other words, 0 as the *result* of 0^n for all non-zero n is not the *product* of a specified but impossible multiplication. Instead, it is the *count* of groups containing 1 units, which is 0.

Postscript 2 – the (non-)effect of the author's view of zero on long multiplication, etc.: How does my view of zero as an operation-blocker affect long multiplication or long division or matrix algebra? The short answer: it doesn't affect them at all. For instance, observe what happens when you multiply 103 by 700 using long multiplication:

```
    1 0 3
    7 0 0
    0 0 0
   0 0 0
  7 2 1
  7 2 1 0 0
```

0 * 3—the result is 0, reflecting that there are no units to count.

0 * 0—the result is 0, reflecting that there are no groups of 10 to count.

0 * 1—the result is 0, reflecting that there are no groups of 100 to count.

[Another row, indented one space to the left, with the same results.]

7 * 3—the result is a product of 21. Write 1, carry 2.

7 * 0—the result is a count of 2 (carried from the previous result).

7 * 1 = 7.

Even though *attempted multiplication* of, and by, 0 does not yield a product, it results in a *count* of 0 (meaning that there *aren't any* units, no groups of 10, etc. to count), and that result is placed in the appropriate slot in the long multiplication, as is conventionally done. And even though *attempted addition* to 0 does not yield a sum, it results in a *count* of the non-zero number(s).

So, long multiplication is not affected by zero's being an operation-blocker. The reason is that the *counting* involved in doing long multiplication is not affected by zero as an operation-blocker. *Zero does not block counting, only other operations based upon counting.*

For much the same reasons, conventionally accepted equations from trigonometry, calculus, and physics are re-interpreted, but otherwise unchanged.

E.g., $y = -x^2 + 1$ is a parabola with maximum of $y = 1$ for $x = 0$ and $y = 0$ for $x = -1$ and $x = +1$. The calculation for for $x = 0$, for instance, is $y = -(0)^2 + 1$, which is interpreted as the sum of (–1 multiplied by two factors of 0) and 1.

Since the factors of 0 block the multiplication, the result is 0, and this blocks the *addition* of 1. However, although the 1 is not added, *it is counted*. This is why the result is that $y = 1$.

Postscript 3 – mercury, money, and metaphysics: It is a sad measure of how divorced contemporary mathematics is from the real world that academic and professional mathematicians – and intellectuals in general – have such a hard time realizing and admitting how "fishy" it sounds to treat zero as a quantity, rather than the absence of a quantity.

In fact, one of the most frequent points I hear in rebuttal is the old canard about how our standard ways of measuring temperature "prove" that zero is a real quantity, and that my claim that zero is an operation-blocker thus cannot be correct.

The first thing that must be noted is that "zero degrees" is used in *two* distinct ways in the Centigrade/Celsius and Fahrenheit scales, and that the one used in the attempted debunking of my model does *not* designate an absence of temperature, while the one that *does* designate an absence of temperature (change) does not pose a challenge to my model.

Suppose, for instance, that it's 32 degrees Fahrenheit at 7 am, and that at 10 am it has warmed up to 42 degrees. That is a 10 degree increase in temperature, the number of degrees change determined by subtracting the first amount from the second.

Now, suppose that at 8 am, the temperature has not yet changed since 7 am. There is thus nothing to subtract, because the temperature remained constant. Now, suppose that by 5 pm, the temperature, which was 42 degrees at 10 am, has cooled back down to 32 degrees.

In this case, while there was not any *net* change in temperature, there *were* two changes of opposite and equal magnitude, and they canceled each other out. These are two very different ways of there not being a net change in temperature – one being a complete absence of change of temperature(as between 7 am and 8 am), the other the result of two opposing temperature changes offsetting each other (as between 7 am or 8 am on the one hand and 5 pm on the other).

We know, of course, that the Fahrenheit scale's temperature of 32 degrees is the Celsius scale's temperature of 0 degrees. They are both defined in terms of the melting/freezing point of water.

However, this does not mean that 0 degrees Celsius is an absence of temperature. It is just an arbitrary point, treating temperatures *below* the melting/freezing point of water as represented by numbers opposite in kind from the numbers representing temperatures *above* the melting/freezing point of water.

The equally arbitrary use by the Fahrenheit scale of 32 to designate the melting/freezing point of water treats the first 32 degrees of temperature below that point as being represented by numbers of the *same* kind as the numbers representing temperatures above that point. Fahrenheit has its own "zero" 32 degrees of temperature below water's melting/freezing point.

But the question is: how do we measure temperature changes from the Fahrenheit or Celsius zero points?

Suppose the temperature rises from 0 to 30. On my model, you cannot subtract 0 from 30 to find the difference. If you have 30 temperature degrees, and you try to subtract 0 degrees from it, you find that you don't have anything to subtract, so the count of degrees is unchanged from 30. We know, however, that the temperature increased by 30 degree units.

Alternatively, you could note that you began with 0 degrees, and that you want to identify how many degrees were added to 0 to get to 30. Well, with 0 degrees, you don't have anything to add the 30 to, so you just have 30 degrees, as if that is what you had started with. We know, however, that the temperature increased by 30 degree units.

What's important to realize, though, is that temperature calculations involving a zero-point don't have the same metaphysical significance as, for instance, money calculations involving the zero-point in regard to assets and liabilities, or profit and loss. For the latter, "zero" truly *does* mean there isn't any money there (because increases and decreases in money canceled each other out).

Most people can see, by inspection, how much it warms up from 0 degrees to a higher temperature. They understand it in terms of *adding the higher temperature to 0 degrees*, not subtracting 0 degrees from the higher temperature.

In other words, people understand the amount the temperature has warmed up from 0 degrees as being fully, explicitly expressed by the *count* of degrees in the warmer temperature – just as the number of chairs in a previously empty room is fully, explicitly expressed by the count of chairs in the no-longer-empty room.

In this way, they know, at least implicitly, that subtraction of zero from that count is unnecessary. (And, I maintain, impossible).

Chapter 8: More Ado about Nothing: The inconvenient Fiction of "Empty Sets"

> [T]he metaphysical meaning of "nothing" is non-existence, the literal void, the blank, the zero...A zero is not given the same status as an entity; the nonexistent is not considered on a par with the existent; and nothing is not taken as equal to a something. [Nathaniel Branden, *The Vision of Ayn Rand*, pp. 28, 68]

> [T]he *Reification of the Zero* [is a fallacy which] consists of regarding "nothing" as a *thing*, as a special, different kind of *existent*... ["Nothing"] is strictly a relative concept. It pertains to the absence of some kind of concrete...You can say "I have nothing in my pocket." That doesn't mean you have an entity called "nothing" in your pocket. You do not have any of the objects that could conceivably be there, such as handkerchiefs, money, gloves, or whatever. "Nothing" is strictly a concept relative to some existential concretes whose absence you denote in this form [Ayn Rand, *Introduction to Objectivist Epistemology*, p. 60, 149]

Modern logic and mathematics declare that there is something called the "empty set," which is considered to be either *a* set, or *the* set, that has no members, and which is a subset of every set. This is a foundational notion in mathematical logic, taken as an axiom of set theory which has been taught not just to college students, but to children in the early elementary grades. Many (and I am one of them) suspect that it is not a coincidence that mathematics illiteracy is rampant among the last couple of generations of school graduates.

In this chapter, I will explore the two main difficulties with this notion of the empty set, namely:

(1) It flies in the face of the Objectivist insight that existence exists and *only* existence exists, that non-existence is not some special kind of thing that exists, but only the absence of something or other that *does* exist.

(2) It is logically incoherent and self-contradictory.

Let's begin to resolve this deep clash between rival perspectives on the foundations of mathematics by examining the seemingly innocuous concept of the "complement" of a set, which is defined in relation to some larger third set, of which a set and its complement are both "subsets." The set and its complement together non-overlappingly comprise the total membership of the larger third set.

For instance, in regard to a set of six apples, the set comprised by two of those apples is the complement of the set comprised by the other four of those apples. There is no problem understanding the meaning of "complement" here, nor of the union of a set and its complement in relation to a larger whole.

However, the fact that a set and its complement are both subsets of a *larger* whole creates an insoluble logical barrier for considering the "empty" set as the complement of any non-empty set.

To "complement" means to *add to* something *in order to make a larger whole*. But since the non-empty set cannot become part of a larger whole if we do not add anything to it, then, by definition, any purported empty set, which cannot add anything to another, non-empty set, cannot logically be said to "complement" that set.

For example, the six apples *already* are a whole six apples, and you cannot meaningfully add *zero* apples in order to make the six apples a whole, because they already *are* a whole. Zero apples is (are?) *no part* of six apples, and thus not only is zero apples *no subset* of six apples, it is *no complement* of six apples.

Thus, it is a misnomer to speak of the *union* of something and nothing, because, for instance, you cannot meaningfully speak of the union of six apples and zero apples. You are not finding the union of *anything* with the set of six apples, because the set of six apples is already a set of six apples.

I don't see how you can escape the implication that the notation expressing a union of the null set with another set simply means that *the operation of set union is not performed*. In other words, it seems inescapable to conclude that the *null set is an operation-blocker for set union*.

Chapter 8 – The Inconvenient Fiction of "Empty Sets"

Suppose one disputes this claim with the following example:

$\{a_1, a_2, a_3, a_4, a_5, a_6\} \cup [0] = \{a_1, a_2, a_3, a_4, a_5, a_6\}$.

Doesn't this clearly state the union of the first set with the empty set? No. There already is a unity in the original set $\{a_1, a_2, a_3, a_4, a_5, a_6\}$. No new unity is formed by the alleged union of this set with { }.

If I say that I have *united* something with nothing, what I'm really saying is that have *not* united something with *anything*. I have not performed an operation of union.

Granted, it *appears* that performing an operation of union of something with *nothing* is indistinguishable in its effects with *not* performing an operation of union of something with anything. However, "uniting something with nothing" is not an operation. It is the *absence* of an operation.

Now, suppose we consider a more complex example:

$\{\{\{a_7, a_8, a_9\}\} \cap \{a_{10}, a_{11}, a_{12}\}\} \cup \{a_1, a_2, a_3, a_4, a_5, a_6\}\}$

It looks rather formidable, but actually it's pretty straightforward, whether in this form or as distributed.

In this form, the intersection of the first two sets is the "empty set." (They have no common members.) The 'union" of the empty set and the third set is simply the third set, $\{a_1, a_2, a_3, a_4, a_5, a_6\}$, because there is nothing to unite the third set with, so nothing is united with it.

As distributed, the union of the first and third sets is $\{a_1, a_2, a_3, a_4, a_5, a_6, a_7, a_8, a_9\}$, and the union of the second and third sets is $\{a_1, a_2, a_3, a_4, a_5, a_6, a_{10}, a_{11}, a_{12}\}$. The intersection of these unions (the members in common between the two unions) is again simply the third set.

As an aside, here is a parallel from arithmetic: in the expressions "0 + 5" and "5 + 0," there is no operation being performed. The meaning of "0 + 5" is: "5 not being added to any number," which is simply "5." The meaning of "5 + 0" is: "5 not being added to by any number," which again is simply: "5."

Similarly, the union of an empty set with a non-empty set is the same as the union of that non-empty set with the empty set. In each case, there is no such thing, for there is no such operation being performed in either case.

So, in the former case, the real meaning of the result of the supposed operation is: the (non-empty) set *not being united* with any set." In the latter, the meaning of the result of the supposed operation is: the (non-empty) set with *no operation of set union being performed* on it by any set.

In each case, numbers and sets, it is as if the number and non-empty set had been stated with a single symbol. As Kant might have said (in a much different context, in his *Critique of Pure Reason*, where he was denying existence as a predicate), *nothing has been added to* (or *united with*) the number or non-empty set.

(Kant, being pre-modern logic and pre-modern mathematics in his thinking, would have used the term "nothing" not like "some empty kind of thing," but like "the absence of anything.")

You don't add non-existence to existence, and you don't add existence to non-existence. Similarly, you don't unite non-existence with existence, and you don't unite existence with non-existence.

Trying to do any of these things is a fruitless, meaningless, empty gesture. You get what you started with, which means you didn't do anything except wave your arms or flap your jaw.

Now, the application of this perspective to a proper interpretation of the ontological meaning of the identity element in addition is this: no operation of addition is being performed. Again, when we say nothing has been added, we mean not that some special kind of thing called "nothing" *has* been added, but that something *hasn't* been added.

It's only in a functional or verbal sense that "not adding anything to something" or "not adding something to anything" is equivalent to "adding nothing to something" or "adding something to nothing." Ontologically, they are as different as...existence and non-existence!

(Here we are talking about the difference between nothing *being* added or added to vs. something *not being* added or added to.)

However, just to be clear, I'm *not* criticizing the mathematical formalism of "$x + 0 = x$" or advocating that we abandon it. Not only would that be incorrect and unnecessary, it would be disastrous in its practical consequences. It would completely hamstring mathematical operations and the myriad practical and technological things that depend on them.

I wouldn't do this any more than I would deny that syllogisms are formally correct or incorrect, regardless of whether the terms in their premises or conclusions have actual referents. Formalism in mathematics, as in logic, is just fine. It's the *standard ontological interpretation* of that formalism I'm saying is not correct.

Specifically, what I'm saying is that when you attempt to perform the operation of addition with zero, nothing happens – and that is not because you are *adding nothing* (0) to x or adding x *to nothing* (0), but because you are *not adding anything to x* and *not adding x to anything*. Nothing happens, because there is no operation.

Mathematical and logical formalisms that *work* in the real world didn't just fall out of the sky or ooze out of someone's overactive imagination or some bodily sphincter. They were *abstracted from real connections* in the world.

In the case of addition, we know that when you don't increase a given quantity by some specific quantity, that given quantity remains unchanged. The count of that given quantity is not extended with an additional count. And when you don't have any specific quantity to start with, then there isn't anything to be changed by the introduction of some specific quantity. The count of that given quantity is not an extension from a pre-existing count.

In all other cases, the result is something different from either of the two quantities on the left side of the equation, because the first is *something* that is *increased* by *something else*. You count the units specified by the first number, then continue the sequence by counting the additional units specified by the second number.

When either number is zero, there is no counting beyond the units specified by the non-zero number, so there is no operation of addition being performed. The additive identity prevails because, when there is only one number involved, that number is itself. In all such cases, the only operation performed is *the counting of that one number*.

Zero does not block counting. It merely blocks the extension of a count, when there isn't anything to extend or to extend from!

All other addition facts are connections abstracted from reality that must be learned by examination, study, and memorization, and checked and re-checked if necessary. The additive identity, however, is a fundamental, *axiomatic* connection to reality that is seen by direct observation and grasped as necessarily true by realizing that to deny it leads to a contradiction.

Mathematical formalisms can indeed generate logically valid *results*, even when zero is involved. Something analogous happens in logically valid syllogisms using false statements or even nonsense terms.

Yes, the middle term acts as the linking term between the terms that are the subject and predicate of the conclusion. However, the logical link in a syllogism does not always point to a real connection between concepts, just as the plus sign for addition does not always point to a real operation of addition being performed.

Just as the operation of addition is abstracted from an actual process of counting the total number of units referred to by two or more numerals, so is the operation of syllogistic inference abstracted from an actual process of identifying a causal connection between two facts,.

The formal structure of causal connections is universal and abstractable from reality. This is why it can be used with *any* propositional content to generate a formal connection.

When the premises are true, then the formal connection also corresponds to reality. But when the premises are false, then it's, as Rand would say, "deuces wild" – or as our computer mavens might say, "garbage in, garbage out."

As an illustration of this principle, consider these two syllogisms, both with a false premise and a valid logical structure:

All cows are mammals. (True)

All mammals are creatures that fly. (False)

Therefore, all cows are creatures that fly. (False)

All bats are mammals. (True)

All mammals are creatures that fly. (False)

Therefore, all bats are creatures that fly. (True)

In each case, the middle term, "mammals," purports to be the *cause* or *reason* in reality of the fact that all cows, or all bats, are creatures that fly. In the first case, the conclusion is false. In the second case, it's true – but it's only true *by accident*. (GIGO.)

Yes, some garbage *just happens* to be true. But as the conclusion of a syllogism with false premises, it's not knowledge, just a true proposition you know to be true *for reasons independent of the unsound syllogism*.

I mention all of this, not to argue that syllogisms are just like addition, but to point out that valid formalisms have a source in the real world, and that empty or false *uses* of those formalisms have to be *understood* and not casually manipulated as though they were arbitrary, logical tinker toys.

Let me express this point about mathematical formalisms another way, for those who like to remind us that zero is the "additive identity." The *reason* that zero is the additive identity is that "$x + 0 = x$" is the equivalent of saying "$x = x$." If you start with x and you do not perform any operation of add something to x, then x remains what it is, instead of becoming something different!

Similarly, "$0 + x = x$" is also the equivalent of saying "$x = x$," in this case because if you start with the absence of anything, you are in fact *not starting with anything* until you introduce x, at which point you are *starting with x*. Again, there is no operation of addition being performed, and x

again remains what it is, instead of becoming something different by being added to something.

Well, because of the additive identity of zero, it's certainly true that the equation "$x + 0 = x$" is necessarily correct. But this is true, regardless of what units are referred to by "x" – *and even if it does not refer to any units at all!*

If $x = 0$ (i.e., if there aren't any units referred to by "x"), then it is indeed true that "$0 + 0 = 0$" – *not* because you are *adding zero* to zero, however, but because you are *not adding anything* to zero. In other words, when $x = 0$, "$0 + 0 = 0$," *because* "$0 = 0$." Again, the true ontological interpretation of zero as the additive identity is clear to see.

In order to bridge back to the "empty set" focus of this chapter, let's consider a real-world example. Suppose I want to find out how many pairs of socks I have in my two sock drawers. I reach in one drawer and pull out 12 pairs. I reach in another drawer and do not find any socks. Now, have I performed an operation of starting with 12 pairs of socks and adding 0 pairs of socks to it?

No. While I can actually count the 12 pairs of socks that are in one drawer, I am not extending the count to include the 0 pairs of socks in the other drawer.

Let's walk through the process: "OK, I'm reaching in my upper sock drawer, and I feel some pairs of socks in there, so I grab them and pull them out. Then I count all the pairs of socks I have pulled up and find that there are 12 pairs of socks in total.

"Now, I'm reaching in my lower sock drawer, to see if there are any pairs of socks that I might add to the 12 pairs of socks from my upper sock drawer. Aw shucks, I don't feel any socks there.

"Hmmm, that means that there aren't any pairs of socks to add to the 12 pairs of socks from my upper sock drawer. So, I won't perform any operation of addition – I won't extend the count of pairs of socks beyond the 12 pairs I found in the upper sock drawer, since there aren't any pairs to add or count."

Chapter 8 – The Inconvenient Fiction of "Empty Sets"

But what if we started with that lower, empty sock drawer. Now toss 6 pairs of socks into that drawer. How many pairs of socks in the drawer? Six. I have just added six pairs of socks to a set of non-existent socks, haven't I?

No, there is no set, empty or otherwise, to add them to. There was only the empty drawer that I placed them in. Again, a set is not like a drawer, or any other kind of container, empty of things or filled with things. It is just a collection of things. If there are no things, there is no collection or set.

The truly bizarre implication of the empty set theory is that *anywhere* and *everywhere* that I *might* place those 6 pairs of socks would already contain "a set of non-existent socks." Furthermore, it still would contain that set of non-existent socks, even when the existent socks were placed there, too!

The universe, by this view, is through-and-through empty sets, waiting for the *possible* addition of six pairs of socks, ten thousand pairs of socks, three horses, twelve bogus mathematical-theoretical arguments for the existence of empty sets, whatever. (Oh, and don't let me forget: a partridge and a pear tree!)

To me, this all smacks of idealism. John Stuart Mill proposed the rather fishy definition of "matter" as: the permanent possibility of sensation. The present view of the universe, similarly, is the rather bizarre notion: the permanent possibility of non-empty sets.

A set is not like a room or a cup or a sock drawer. A set is a collection of objects. The room or cup or sock drawer or area on a table or on the ground where the pennies are placed is not a set before the pennies are placed there. It is just a room, cup, etc.

If I showed a friend my empty china cabinet and said: "What do you think of my set of china?" he would rightly ask: "What set of china?"

A modern logician might reply: "You must be blind – or at least mathematically unenlightened. It's an empty set of china." To which my friend would rightly respond: "Oh, yeah, right next to the empty set of sane ideas that are coming out of your mouth right now."

Empty rooms and empty sock drawers and empty china cabinets and empty containers exist, but *empty sets do not exist*. A set is not a container. A set is a collection of things that results from the *mental* collecting of some specified things. If there are no things of a particular kind to collect, then there is no collecting, no collection, and *no set*.

Modern logic's embracing of the empty set is very similar to, and just as destructive philosophically as, the idea that the universe could somehow be devoid of things that exist. The universe is not a place or a container that holds all of the things that exist. It *just is* the sum total of all those things. If nothing existed, there would be no sum total and no universe.

But even to say "if" in this case is a mistake. The whole notion that the universe's existence is "radically contingent," that it's metaphysically possible that nothing could have existed, is often argued in terms of the universe being a set or a "sum total" of things that exist, and that that "set" could have been "empty." Because after all, we can have "empty sets," don't you know.

Now, suppose we take the perspective that sets are not physical collections, but instead are mental collections resulting from neural processes taking place in our brains and nervous systems. This does not alter anything in our analysis to this point.

If a set is merely a non-physical entity (?) or mental collection of items, and the set is something (whatever-it-is) that is generated by physical brain processes, it still has a nature. It still has contents, whether or not those contents correspond to physical reality.

To carry out a process of mentally collecting items and forming a set, the brain has to operate on the products of other brain products. This is so, whether those products are the result of currently perceived items, or of items that have been remembered, or whatever.

Your brain can produce a mental image of a green triangle, and so it can also produce a set containing one or more such imaginary items. Greenness and three-sidedness *can* coexist in reality, even if they in fact *may not*.

Chapter 8 – The Inconvenient Fiction of "Empty Sets"

Granted, your brain cannot produce a mental image of a four-sided triangle, because four-sidedness and three-sidedness do not and cannot coexist in reality – and therefore such a thing is not even imaginary but, instead, a full contradiction. Nevertheless, the set of such impossible items would not be empty, but full of arbitrary posits of that kind.

These two sets are equally real (equally brain-generated, that is), but the latter is no more empty than the former. The difference is that the mentally constructed (brain generated) contents of the former may but need not exist in reality, while the mentally constructed (brain generated) contents of the latter cannot exist in reality.

Neither of these sets is an *empty* set, however, unless we arbitrarily define "set" as that which contains *real* (mind-independent) items – and I don't know anyone in or out of the world of modern logic who wants to claim that. But even then, we would be ruling out empty sets, since something which does not contain real items is by definition *not* a set.

Modern logicians, of course, disagree strenuously with this view of empty sets. They believe that the empty set is as much a real thing as a non-empty set. If a non-empty set is a collection that has members, then an empty set is a collection that does not have any members.

To grasp the illogic of this notion, consider the parallel to ethics. If morality is a code of values to guide your actions in living your life, then an empty morality would be a code with no specific values to guide your actions in living.

Think of your moral code after you have betrayed all your values. The code of values in your morality is empty. If you find this nonsensical, you are not alone.

Let's return again to the sock drawer. If my socks are all in the washing machine or my suitcase, then my sock drawer is empty. There is no "empty set" of socks in that drawer. It is an empty sock drawer. It does not *contain* an "empty set" of socks.

The set of socks is the group of socks, *wherever they are*, whether they are in some container or not. The chest of drawers is no more an empty set of socks than the washing machine is a full set of socks. Those containers are just places where the socks are or aren't. The socks themselves are the set, and wherever they aren't, there is no set of socks, full or empty.

A set is *not* like a bucket or a room or a sock drawer or a silverware chest – a container that either has things in it or does not. A set *is* a totality of *things gathered*, either mentally or physically, whether or not they are also *in* a container of some kind.

A silverware chest with no knives, forks, or spoons in it is not an "empty silverware set" and does not *contain* an "empty silverware set," any more than it is, or contains, an "empty set of square circles." The silverware set *is* the collected totality of knives, forks, or spoons – *wherever they are* – not the thing that they are, or are not, placed in at any given point in time.

A concept, which is a set of like things, *is* those things *as the mind/brain holds them to be a group of similars*. We speak of the "content" of concepts, but that does not mean that concepts are empty forms or containers into which we pour mental contents of one kind or another.

Concepts *are* the things that are *mentally formed into* a single unit, the group of (i.e., grouped together) similar things. The form is the grouped content, not some prior existing container into which the grouped content is poured.

The same is true for sets, but only more generally, since sets are not necessarily a grouping of similars, just a grouping of things selected for attention on *some* basis or other.

Now, it might be helpful to clarify whether we are talking about *an* empty set or *the* empty set – although I'm not sure what difference the answer would make. The whole concept has seemed incredibly bizarre and nonsensical to me from the first time I heard about it.

First, consider the fact that, whether or not my (designated) sock drawer is empty at the moment, set theory tells us that it should contain an infinite number of empty sets.

Now, note: this does not include merely the (temporarily) empty set of my socks which presently may be in the clothes hamper, the clothes dryer, or my suitcase, but also the empty set of my wife's socks, the empty set of green triangles, the empty set of round triangles, etc. ad infinitum.

Even worse, it seems to follow that this infinity of empty sets is *everywhere* that the things they don't contain (!) *aren't* – as well as everywhere that they *are*! (The empty set is a subset of all sets.) Not only in empty spaces, but even in spaces that are completely filled with other things!

Doesn't this seem cognitively bizarre – populating the world – both its empty areas and its full areas—with an infinity of nothings? What is the cognitive purpose or value of doing this?

Secondly, is "the empty set" a master set that contains all of the particular empty sets? If so, is *it* empty, as its name suggests? In which case, how can it contain particular empty sets? I.e., if all the specific empty sets are things that "the empty set" contains as subsets, how can it be empty?

On the other hand, if "the empty set" is *not* an über-set containing all of the particular empty sets, then what is it? What is its relation to the infinity of specific empty sets, and what possible use does it fulfill?

I think that there is *no* good justification for the empty set – by which I mean that there isn't *any* good justification for the empty set. (I say "isn't any," rather than "is no," because strictly speaking, *"no good justification"* for the empty set is not *something* that exists!)

Suppose we argue that there is not an infinity of empty sets, here, there, and everywhere, but just one and only one empty set. How might we understand or justify this? It would seem that we would have to somehow show that any two presumably distinct empty sets are really not distinguishable.

First, we note that two sets that have exactly the same members are equal. But having exactly the same members is precisely equivalent to having exactly the same non-members. If *everything* is a non-member of each set, then they are not only equal, but empty sets.

Thus, it is a misnomer to think of there being two (or more) empty sets. They are all the same, indistinguishable – hence they are all *one and the same*.

On this view, it is the one and only "empty set" that is everywhere. Not just the "empty set" of my socks in my presently empty sock drawer, nor just the "empty set" of my wife's socks in my presently empty sock drawer, nor just the "empty set" of blue triangles or round triangles in my presently empty sock drawer.

No, the one and only "empty set" includes not only *all* of those, but also *infinitely more* that are all *that same, one and only* set that is in my sock drawer when my socks *are* there! And *everywhere else* in the universe!

Nothing – i.e., the "empty set" – is everywhere! It is everywhere that things are – and everywhere that they aren't! I couldn't agree – and disagree – more.

Now, I do agree that everything (that exists) is somewhere in particular, and in that sense there is not any thing (that exists) that is everywhere in general. But that is radically different from saying that "nothing" in the sense of *that which does NOT exist* IS everywhere – which is what set theory's "empty set" construct implies.

My socks can and are absent from nearly everywhere in the universe. But that does not mean that their "non-existence" *is* here, there, and (nearly) everywhere. It only means that *they aren't* in those places! And in general, and without exception, each and every thing that exists is only where it is, and not where it isn't.

So, unlike 5's, of which there is the abstraction and the many specific instances of 5 things, there isn't one "empty set," and there aren't *many* "empty sets." There is not any "nothing." There is only something that is.

This is why, when we say "there is nothing in my sock drawer," we aren't saying "there is an empty set" in that drawer. When we say that the drawer "contains nothing," we aren't saying that it's performing a function of containing the absence of something.

Because anything which my empty sock drawer *might* contain is not there, but somewhere else, what we're actually saying is that it *doesn't* contain *anything*.

I hope the reader can see what a vast difference there is between these two perspectives. One of them leads to Reification of the Zero, and the other affirms the Primacy of Existence.

Now, having said this, I must assure the reader that I do not have a problem with the word or concept "zero." Zero, properly used, is not taboo, and I don't have "zerophobia." To the contrary, I think I have a very clear understanding of and respect for zero – "zerophilia"!

I love the concept of "zero," for what it is and for what it is able to do. I just don't think you can use zero to *do* things to other numbers, nor to use other numbers to do things to zero. It is simply a place-holder for what is not there, or a device to indicate what is *not* done *to* something else – or what is not done to it *by* something else.

When you state that "kinetic energy is zero" or "the number of chairs in the room is zero," you cannot be stating a measurement or count. You cannot count or measure what is not there. All you can do is observe *that* it isn't there, and designate that absence by "zero."

Specifically, the most you can possibly mean by "zero kinetic energy" or "zero chairs in the room" or "added zero chairs to (or subtracted zero chairs from) the room" is that there *isn't any* kinetic energy, or that there *aren't any* chairs in the room, or that you *haven't added any* chairs to, or *subtracted any* chairs from, the room.

Perhaps a "real-life" example would help clarify this point: I can assure you that my dear wife would much prefer that I assure her that I *haven't* slept with *any* other women, than to tell her that I *have* slept with *zero* women.

(Not only would she find the latter a bit odd, but also probably disquieting. For one thing, if you want to reassure someone about what you *haven't* done, you probably shouldn't start out by saying that you *have* done something and then quantify it as "zero.")

Another relevant example for many of the people I have discussed this matter with would be: "Ayn Rand had zero children." From what I've just said, it should be clear that this would be better expressed as: "Ayn Rand did not have any children."

(In standard propositional form – which is best for doing deductive logic – I would replace "had" with a form of "to be," and I would express the predicate as being in the same category (entity or person) as the subject. Viz., "Ayn Rand was not a person having children," or: "Ayn Rand was not a parent." The latter is actually verbally simpler than "Ayn Rand did not have any children.")

Now, you can't *have* zero children. You can't *have* a "complete absence of any quantity or magnitude" of *something*. Instead, in such a situation, you simply *don't have* "any quantity or magnitude" of something.

Yet, some people profess to see no difference between saying: "Ayn Rand had zero (or no) children," and: "Ayn Rand did not have any children." This is very much like the people who don't see any difference between saying: "Atheists believe in no God," and saying: "Atheists do not believe in God."

In each case, a very important distinction is wiped out between a positive action ("having" or "believing") with no object and an absence of an action with an object.

Blurring or erasing this distinction may be acceptable in nursery rhymes – "Jack Sprat could eat no fat, his wife could eat no lean," "And so her poor dog had none" – but it has a very pernicious effect in logic, philosophy, and propositional speech in general. Oh, yes…and mathematics.

This careless and ill-considered reification of zero is very pernicious and has the potential to infect and corrupt every one of your thoughts – if you let it. So, don't.

Postscript – empty sets as operation-blockers: This topic reminds me of the old saw about evidence and justification: absence of evidence is *not* evidence of absence. Nothing is not something. In other words, I think we're onto something here – and it's not nothing!

Chapter 8 – The Inconvenient Fiction of "Empty Sets"

What we are seeing here is nothing less than the deep, ontological meaning of operations conventionally taken to involve zero or empty sets. The operations are actually being specified as not having been performed!

In this way, a number of mathematical and logical expressions conventionally regarded as arbitrary premises in order to build a system of inference can instead be seen as specifying that zero and empty sets are operation-blockers.

To be sure, their roles as operation-blockers does not prevent them from being the *results* of arithmetic or set operations. Just as zero can be the result of an arithmetic operation, so can the empty set be the result of an operation on sets. In particular, arriving at the empty set as the result of intersecting two incompatible, non-overlapping sets is no problem.

For instance, the intersection of {1, 3, 5} and {2, 4, 6} is {0}.

However, the empty set cannot be conjoined with a non-empty set, or a non-empty set with an empty set, any more than 0 can be added to x, or x added to 0. You cannot add 5 chairs to 0 chairs, for there isn't anything to add the 5 chairs to. And you can't conjoin a non-empty set with an empty set, for there isn't anything to conjoin the non-empty set's members with.

Similarly, you can't add 0 chairs to 5 chairs, for there isn't anything to add to the 5 chairs. And you can't conjoin an empty set with a non-empty set, for there isn't anything to conjoin with the non-empty set's members.

Zero and empty sets are operation blockers more fundamentally because their base, the concept of *nothing* is also an operation-blocker. Nothing does not exist. You can't get inside it, outside of it, around it, underneath it, period.

All that exists is Existence, and Existence is *all* that exists. It is a complete sum total. It cannot have a complement, because there isn't anything you can add to it. And you especially can't add Nothing to it, because Nothing isn't anything at all, let alone anything that can be added to something that exists.

"Nothing" or "non-existence" only has meaning in relation to some specific thing that might or might not exist, but even then, it's an operation-blocker. If you look into a room that contains a table and chair, and someone asks you what you see, your perceptual mechanism finds the two objects to lock onto, and you report, "I see a table and chair."

But if you look into an empty room, and someone asks you what you see, how do you reply? Would you say, "I see nothing there"? Perhaps, but what you are really saying is, "I *don't* see *anything* there." You are not *seeing nothing*. You are *not* seeing *anything* (except a room).

The absence of anything in the room is an operation-blocker – in this case, the operation of perceiving entities. There isn't anything for your perceptual mechanism to lock onto (except for the room itself), so your entity-perceiving function is blocked.

So, Existence as the set or sum total of everything that exists cannot have a complement. Existence as a sum total *must* exist. It cannot go out of existence, so it has no "opposite" either – no whatever-it-is that there would be if Existence stopped existing (because it can't).

To repeat what I said previously: you can't add non-existence to existence, and you can't unite non-existence with existence. Trying to do so is a fruitless, meaningless exercise. You get what you started with, which means that you didn't do anything except wave your arms or flap your jaw.

Postscript 2 – are numbers merely imaginary? Somewhat paradoxically, some proponents of modern mathematics argue not only that the empty set is no less real than non-empty sets, but also that numbers designating some actual quantity are no more real than zero which does not designate any actual quantity.

A typical comment: no matter how far and wide you cast your net in the universe, you will never find an integer existing as a material, physical entity.

I find this rather amusing, and so would the other two people currently residing in my house. Someone asked to find an example of "three" in my house could immediately point to me, my wife, and my daughter – or our three computers – or our three placemats – or any three objects.

An integer is not *an* entity, but *some* (a group of) entities, however many that integer is. It's true that there are no integers (or groups) apart from entities – but it's also true that there are no entities that are not some integer in number, whether as individuals (one) or as part of a number of individuals.

I am a proud individual (one) *and* a member of an indefinitely large number of groups of different numbers of things. That is *what* we "construct" the numbers we use mentally *from*.

Yet, the irrealists in regard to number say that integers exist only in our heads, not independently in reality. In this respect, they say, the real numbers are just as imaginary as the "imaginary numbers." If there were no intelligent beings in the universe, there would be no numbers.

Correction: there would be no *numbered* (counted) things. But there are three human beings in my house now, whether or not any one numbers (counts) them. That *is* how many human beings *there are* in my house, and that remains true (an "objective" fact), whether or not there is anyone to perceive and count those three human beings.

Even if every intelligent being in the cosmos suddenly died right now, there would still be three dead humans in my house right now. To claim otherwise is confusing number (a quantity) with count (a measurement of quantity).

Number and quantity are intrinsic to reality. There are no things that do not have *some* number and quantity apart from human or other awareness of it. By contrast, count and measurement are "objective" – in the Randian sense of: the product of a consciousness being aware of number and quantity.

Thus, confusing number and quantity with count and measurement is a conflation of the intrinsic with the objective, just as surely as if we were to ignore attributes, which exist independent of consciousness, and focus only on qualities, which are the form in which we are aware of independently existing attributes.

Certainly qualities and count and measurement do not exist apart from sentient beings. But attributes and number and quantity do.

If a tree fell in the forest and crushed three deaf humans, but there were no one there to hear it and to count the three dead humans, would it still have made a sound and would there still be three dead, deaf humans? Yes. There would have been no experienced sensory quality of sound and no counting of the three dead, deaf humans, that's all.

Postscript 3 – other examples of non-existence as an operation-blocker: I've made quite a bit out of how we use language in referring to the absence of something. I have insisted that it literally makes no sense to speak of doing this or that mathematical operation to nothing, whether in its incarnation as the number "zero" or its incarnation as the "empty set."

This issue is not just a (seemingly) pedantic, abstruse, logical or mathematical concern. It pervades our lives, including in particular the areas of law and religion. It would behoove us to get it right, across the board, since there are situations where the right view is a matter of life and death – not to mention eternal salvation and damnation!

Take the issue of guilt or innocence in the area of legal justice, where "nothing" is involved in two important ways.

First of all, when a person is taken to court and tried for having committed some illegal act, the burden is on the prosecution to prove that the person did in fact commit that act – that there *is* an illegal act that the defendant committed.

The defendant, on the other hand, is not required to "prove a negative," which means that he does not have to prove that he did *not* commit the illegal act. In other words, he does not have to prove the non-existence of a certain illegal act that he committed.

The prosecution is required to demonstrate and prove the *existence* of something. If they fail, the outcome is not that they have "proved nothing," but that they have "failed to prove anything."

Chapter 8 – The Inconvenient Fiction of "Empty Sets"

By the same token, the defendant doesn't have to prove *non-existence*. He is no more capable of "proving nothing" than is the prosecution. All he has to do is point out that they have *not proved anything*.

Secondly, when the prosecution fails to make its case, the defendant is declared "not guilty," which means that he *has not been proven guilty* – that is, that the existence of an illegal act by him has not been proved.

It does *not* mean that he did not in fact commit the act, that there is no illegal act that he committed. In other words, it does not mean that he is *innocent*. He may or may not have committed the act, and thus he may or may not be innocent. All we know from the "not guilty" verdict is that the *existence* of an illegal act by him was *not proved* by the prosecution.

In the field of religion and theology, we have the perennial issue of whether atheism is a religious belief system or the choice not to adopt such a belief system. In other words: is an atheist a person who *believes* that there is *no* god – or simply a person who *does not believe* that there is *any* god?

This distinction is sometimes referred to as "positive atheism" vs. "negative atheism." Positive atheism is the position that there is *no* god – i.e., that there are *zero* gods in the universe. This person is in a lot of trouble with the God of the Old Testament, who said: "The fool in his heart says there is no God."

The negative atheist, by contrast, says that the burden of proof is on the theist to show (give evidence and proof) that there is a god. Failing that, the negative atheist does not say there is *no* god, or that he *believes* that there is no god. Instead he simply says that, without evidence and proof, he does not have any reason to believe in god, so he does not believe in god.

The negative atheist, like the defense in a criminal trial and like Doubting Thomas in the Bible, says: "I'm from Missouri. Show me." It is up to the theist and the criminal prosecution alike to provide reason – evidence and proof – for belief in existence of something, whether a supernatural being or a crime that has been committed by the person being accused of the crime.

So, in law and religion, as in mathematics, non-existence is an operation-stopper. In law, the non-existence of evidence and proof stops the operation of criminal conviction. In religion, the non-existence of evidence and proof stops the operation of belief in a supernatural being. (Or, it should.)

Postscript 4 – Is the "operation blocker" idea just a semantic quibble?
Fred Seddon, in his review of this book (*The Journal of Ayn Rand Studies*, December 2014), said he thinks that the difference between my view of how zero functions in mathematics and the conventional view, as discussed in chapter 7, is a distinction without a difference. If the result is the same, then there is no real difference between saying it's the sum of adding or multiplying with zero and saying that it's the *non-sum* of zero blocking addition and multiplication. (He apparently has the same opinion of my claims about the null set made in this chapter.)

I disagree. The result of attempting to add zero to any number, or to add any number to zero, is not actually a new number which is the *sum* of two numbers, one of which is zero, but simply the *result* of attempting to perform an impossible operation, an operation which is impossible because attempting to use zero blocks the operation from happening. The fact that the rules of arithmetic *stipulate* that this is a legitimate operation only means that people are misled by that stipulation into thinking that attempted addition involving zero is actual addition. But this is no more true than to believe that *not* having *any* coins in one's pocket means the same thing as believing that one has some specific number of coins in one's pocket that is not 1, 2, 3, or any other counting number of coins, but "zero coins." There is no such thing. A supposed process of counting that results in a supposed number called "zero" is *not* counting, but instead the result of *attempting* to count and not finding anything to count! Which is the point of my comments in this chapter as well.

Contrary to Seddon's suggestion, it is *precisely* the idea of "nothing" as a *relative* concept – zero coins or no coins as the *absence* specifically *of coins* – that grounds my view that zero is not a special number of things but the *absence* of *any* number of things. And it is exactly the reification of "nothing" in the form of the *number* zero – treating it as a kind of *thing*, rather than the *absence* of a thing – that is smuggled into the traditional, conventional theory of arithmetic.

Chapter 9: ...and Everything! Thoughts on Induction and Infinity

> An arithmetical sequence extends into infinity, without implying that infinity actually exists; such extension means only that whatever number of units does exist, it is to be included in the same sequence...The concept of "infinity" has a very definite purpose in mathematical calculation, and there it is a concept of method. But that isn't what is meant by the term "infinity" as such. "Infinity" in the metaphysical sense, as something existing in reality, is [an] invalid concept. The concept "infinity," in that sense, means something without identity, something not limited by anything, not definable. [Ayn Rand, *Introduction to Objectivist Epistemology*, pp. 18, 148]

There is a lot of intellectual corruption in modern mathematics, and this chapter is going to focus on one glaring and persistent example of it: the notion that there are "orders of infinity" and, in particular, that there is the same number of counting numbers as there is of even (or odd) numbers.

From a common sense standpoint, this idea is nonsense. However, as Ayn Rand might have said, we don't know that it is nonsense, until we can understand and prove that it is nonsense.

My purpose in pursuing this is not to throw out the accurate but mislabeled or misconceptualized results of modern math. I just want to understand them in the same manner that I understand abstractions more generally, as per Ayn Rand's theory of concepts – and make sure that they fit coherently with the rest of my knowledge of reality.

I've been highly impressed with Ayn Rand Institute lecturer Pat Corvini's lectures on the philosophy of mathematics. Her most recent lectures on the theory of number ("Two, Three, Four and All That," 2008) address the issue of the supposed one-to-one correspondence between infinite "sets" and pretty well demolish the idea.

I'm not going to present Corvini's argument here. She is working on a book, and I invite interested readers to purchase the lectures, which are available as digital downloads, for a pittance from estore.aynrand.org – or to wait for the book. Instead, I am going to share my own current understanding, as it has been enhanced by Corvini's and others' comments.

In my not so humble opinion, the modern claim that the number of even integers is the same as the number of integers, because you can (supposedly) establish a one-to-one correspondence between them, is balderdash, pure and simple.

It is blindingly obvious to anyone who has studied grade school arithmetic that the number of integers between 1 and 10 is ten, and the number of even integers between 1 and 10 is five – and that the same is true for the integers and even integers between 11 and 20, 21 and 30, and so on. The ratio between them of 2:1 remains constant for every observable or conceivable span of ten integers – and why wouldn't it?

How, then, can that ratio suddenly (so to speak) flip from 2:1 to 1:1, when the span of integers becomes infinitely large? There is no rational or empirical reason to think it would. Yet, that is the claim of Georg Cantor and his partisans.

The truth of the matter is that, being infinite sets, the counting numbers (positive integers) and the even numbers are not *actual* sets, but only potential sets. You can count and collect integers until the end of time, and you will never finish the set. There will always be another one, and another, and another...

Thus, infinite sets like the counting numbers and even numbers do not have an actual *number* of members. As a result, not only can the members in each infinite set not actually be counted, but they cannot even be actually *compared* to one another.

When comparing the number of members of two sets, we use either pairing or something based on pairing, which means a *finite* process – i.e., one that has a beginning and an end. *If you begin pairing but do not (and cannot) finish it, then you are not actually pairing, and any conclusion you claim from that truncated process is invalid.*

Chapter 9 – Thoughts on Induction and Infinity

When working with finite sets, you can easily compare the counting and even numbers. It is a process with a beginning and an end, and by using it, you can see that for any finite interval (e.g., 1 – 10, 1 – 100, 1 – 1,000,000,000), the matching operation *always* shows there to be twice as many counting numbers as even numbers.

But when working with infinite sets – e.g., the counting numbers and the even numbers – the pairing operation by which you compare them can never come to an end. Thus, the concept of "equal number of members" or "equinumerosity" cannot even be applied.

So, anyone who thinks that the method of pairing can be applied to a situation with no definite numerosity, and in which the pair operation cannot come to an end...well, that person needs to check his premises.

Think about it for a moment. Regarding the infinity of integers – how can the concept of "set" even apply to such a..."collection"?

By its nature, doesn't "collection" or "set" or "group" (in the everyday, not the mathematical sense) refer to something that is bounded, having a finite number of members? Isn't it illegitimate on the face of it to extend these concepts to the realm of the infinite?

If so, then isn't it just as illegitimate to use the concepts "ratio," "the same as," "twice as many as," "as many as," "equal," "equinumerous," etc. in the context of discussions of the infinite?

For instance, Cantor's advocates argue against using the concept of "ratio" in comparing the size of the sets of counting numbers and even numbers, and would deny our right to say that there is a two-to-one ratio between them.

Yet, they also tell us that there are *the same number of numbers* in each set, because (they and Cantor say) there is a one-to-one correspondence between those members. Isn't this smuggling back in the concept of "ratio"?

Looks like a double standard to me – and in support of a conceptually flawed perspective as well.

But to expand a bit on my earlier point: between 1 and 10, there are twice as many counting numbers as even numbers – between 11 and 20, the same – and so on. There is no evidence and no reason to suspect that this ratio ever changes. It is not a converging series.

Any purported argument that the ratio of counting numbers to even numbers is 1:1 needs to provide something it cannot: evidence or logic.

Logically, such a pairing operation, by definition, cannot come to an end. Empirically, such a pairing operation, in practice, cannot demonstrate even a scintilla of an iota of an indication that the repeatedly observed 2:1 ratio even twitches minutely in the direction of the purported 1:1 ratio. Thus, there no logic or evidence to support the claim for an *actual* one-to-one correspondence between the counting numbers and the even numbers.

Thus, it's obvious to me that:

1) The technique/operation of pairing numbers is derived from concrete reality for the purpose of comparing groups that have a definite numerosity, and therefore is understood as a process that *has a definite end*.

2) Inductively, since for all finite intervals, the number of counting numbers is twice that of the number of even numbers, there is no reason – other than Cantor's arbitrary misapplication of one-to-one pairing – to believe that the number of *all* counting numbers should *not* be twice that of the number of all even numbers, if such totals could even be determined. However…

Not only is counting the number of items in an infinite set impossible, but so is comparing the number of items in *two* such sets. You can match sheep against pebbles and eventually come up with a determination of whether they have the same or a different number, because the process of matching is finite.

But you cannot do this with the counting and even numbers. It is not really a one-to-one matching, because you don't and can't finish it, and each time you check your progress, you see that the matching has not budged from a two-to-one matching and gives no rational indication that it will.

Chapter 9 – Thoughts on Induction and Infinity

In other words, Cantor and his followers have dropped the cognitive context of the matching operation, which is our fundamental tool for evaluating the numerosity of any given, finite, determinate group of things. The notion of a "one-to-one correspondence" between the counting and even numbers is, in Randian terms, a "stolen concept."

Even if the pairing operation *could* come to end, there would be an insurmountable difficulty for Cantor's claim. If each span of ten integers has twice as many integers as even integers, how can the total of even integers over *all* intervals ever average out, so as to be equal to (i.e., in one-to-one correspondence with) the integers?

Presumably, at some point, in order for this to happen, the number of even integers in an interval has to exceed the number of integers in that interval. When does this happen?

Never. At no point.

So, sorry, Cantor and friends. There *must* be twice as many integers as even integers. If it is so for any finite range of integers (granting it has an even number of integers), then it has to be so for an infinite range of integers –any alleged "one-to-one" correspondence notwithstanding.

Here's another way of seeing (through) the flawed thinking behind Cantor's alleged one-to-one correspondence of counting and even numbers.

Suppose I decide to start by matching the first 10 counting numbers to the first 10 even numbers. I note that I have used counting numbers up to 10, but even numbers up to 20.

However, this is misleading. Even though in *matching* the first 10 evens with counting numbers, I only had to use the counting numbers up to 10, in *identifying* the first 10 even numbers, I have had to use the counting numbers up to 20. The first 10 even numbers are *only derivable* from the first 20 counting numbers.

In other words, I have had to *make use of* the first 20 counting numbers *in order to generate* the first 10 even numbers. So, simply staring at the supposed one-to-one correspondence is very deceptive. The second 10 counting numbers are already in the game, even though they're not explicitly laid out in the pairing process.

Inductively, this is true for *any* finite or infinite set of counting and even numbers. Even though in *matching* the first N evens with counting numbers, I only have to use the counting numbers up to N, in *identifying* the first N even numbers, I have to use the counting numbers up to 2N. The first N even numbers are *only derivable from* the first 2N counting numbers.

In other words, I have to *make use of* the first 2N counting numbers *in order to generate* the first N even numbers. So, again, simply staring at the supposed one-to-one correspondence, as this time it trails off into the infinite distance, is very misleading. The second N counting numbers are already there, even though not explicitly laid out in the infinite, one-to-one, pairing process.

The flaw in Cantor's gimmick is intuitively obvious to an intelligent high school student. What is intuited is that there is some smuggling going in, some unacknowledged use of numbers that are then disowned or surreptitiously swept under the rug in the process of analyzing and evaluating the pairing. In other words, if you use the second N counting numbers, and then deny that you are using them, you are committing the fallacy of equivocation.

Or, again in Randian terms, you are committing the stolen concept fallacy: you are using one aspect of the nature of the second N counting numbers in order to deny another aspect of their nature.

Here's how it works: the second N counting numbers are needed in order to generate the first N even numbers, and then they are discarded from the pairing, as though their previously necessary existence did not establish that there were twice as many counting numbers as even numbers.

In even more simple Randian terms, you are biting the mathematical hand that feeds you. You are trying to have your integers and eat them, too.

Chapter 9 – Thoughts on Induction and Infinity

A *proper* correspondence between the counting and even numbers would note that for every even number, there correspond exactly *two* counting numbers, the two needed to *generate* that even number – namely, $2n - 1$ and $2n$.

Specifically, it would make explicit the facts that:

1) In generating the even number 2, you are using the counting numbers 1 and 2.

2) In generating the even number 4, you are using the counting numbers 3 and 4.

3) In generating the even number 6, you are using the counting numbers 5 and 6. (Etc.)

And that what you are *not* doing is using the counting number 2 to generate the even number 4, the counting number 3 to generate the even number 6, etc.

This correspondence keeps the game honest. It acknowledges that *two* counting numbers are needed to generate every even number. It does not pretend that one of them does not exist, in order to make the bogus argument that there are as many even numbers as counting numbers.

It avoids the larcenous gimmick – or incompetent error – that Cantor introduced over a century ago. (I can't help but note the similarity between this issue and the fact that while every child has two parents, it has become fashionable for people to try, in oddly Cantoresque fashion, to pretend that the father does not exist. In fact, am surprised that the radical feminists have not adopted Cantor as their patron saint.)

We can demonstrate this in terms of multiplication by 2, but since multiplying is based on adding, which is based on counting, we shouldn't be too surprised to find that the analysis reduces to an illustration based on the sequence of counting numbers.

1(2) *is* the last number of the sequence 1 2 [stop]

2 (2) *is* the last number of the sequence 1 2 [stop] + the last number of the sequence 1 2 [stop], which in turn *is* the last number of the sequence 1 2 3 4 [stop].

3(2) *is* the last number of the sequence 1 2 3 [stop] + the last number of the sequence 1 2 3 [stop], which in turn *is* the last number of the sequence 1 2 3 4 5 6 [stop].

That is why the correspondence between counting and even numbers has to look like *this*:

Counting numbers: 1 2...3 4...5 6 etc.

Even numbers: 2....4....6 etc.

Not like this Cantorian con-job:

Counting numbers: 1...2...3...4....5....6 etc.

Even numbers: 2...4...6...8...10...12 etc.

In other words, the correspondence between the counting and even numbers is actually a *two-to-one* correspondence. That is why there are twice as many counting numbers as even numbers.

Thus, the alleged one-to-one correspondence of Cantor proves nothing. As I explained previously, if you use multiplication by 2 to generate the evens, it is just a shell game, a rather lame con job on unsuspecting students (and all too many academics).

Rather than multiplying by two, the simpler way of generating the evens is just to sing out every other number in the counting number sequence: (one) TWO (three) FOUR (five) SIX...etc. Then you can see clearly that instead of "getting" the even number four from the counting number two, you get it (in sequence) from the counting number four, just as you "get" the even integer two in sequence from the counting number two.

Similarly, instead of "getting" the even number six from the counting number three, you get it (in sequence) from the counting number six – and the even number eight (in sequence) from the counting number eight. This is the simplest way I have come up with so far of demonstrating how you can see by inspection that you will always have twice as many counting numbers as evens.

In this respect, it seems clear that Cantor's alleged one-to-one correspondence between the even numbers and the counting numbers is actually between the even numbers and *some* of the counting numbers. Namely, the even number 2 corresponds to the counting number 2, the even number 4 corresponds to the counting number 4, etc.

Although you never finish the "mapping" process, if you could, you would find that half of the counting numbers are left over. Obviously, the numerosity of the two infinite sets cannot be the same.

Here is a helpful analogy offered by my friend Merlin Jetton: imagine that we have all the integers in a container. (Obviously, this is impossible, because there are no infinite containers, but this is a thought experiment.)

Now, remove half of the integers, say, the odd ones. Once you have done this (again, merely hypothetically), it is absurd to claim that as many integers remain in the container as before. Yet, that in effect is what Cantor and his supporters do. They are claiming that when the infinite glass is half full, it's full – but also, that when it's full, it's also half full!

Mysterious indeed are the ways of The Infinite. The Infinite Part, Infinite Whole, and the Infinite Double are a Holy, Inscrutable Mystery, not to be questioned or understood by mere mortals, who think in terms of finite parts and wholes. Three in One, One in Three, the Holy Infinite Mathematical...they are and are not the same.

We should not be surprised to find that Cantor was a mystic, who believed his ideas on transfinite numbers to have been communicated to him by God. (See Dauben 2004.) His inscrutable claims about infinite sets make somewhat more sense in light of his philosophical leanings.

Postscript – the intellectual and moral bankruptcy of modern math and science: There are various reasons why no Objectivist has yet been in the forefront of math and science, but misunderstanding Cantor is *not* one of those reasons!

One of the actual reasons is that modern math and science are such an incredible conceptual mess, that it would take decades for one really focused good thinker to undo the errors – to understand the false premises of the mainstreamers' arguments, and to formulate the correct replacement conceptual framework.

Until very recently, most published Randian philosophical work has been done in ethical and political theory. This is not an accident.

Although ethics and politics are logically and hierarchically dependent upon metaphysics and epistemology (and science, among other things), they are also closer to people's experience and concerns, and are experienced by most as being *the* crucial things to understand and get right.

But as radical and clear as the work in these areas by Objectivists (and libertarians) has been, egoists and capitalists are *still* subject to the slings and errors and smears of those who fear and hate those who threaten their altruist/collectivist rackets.

In fact, if anything, the condemnation and ridicule have intensified in recent years, even as the body of work and the number of pro-individualist thinkers and think tanks has proliferated.

This kind of belittlement of those one disagrees with is not just a phenomenon of the ethical-political realm, however. It is just as likely to occur in discussions of math or science as it is politics or morality.

The problem is not just the questionable wisdom of using ridicule and scorn as an *adjunct* to reasoning, in order to spice up and enliven and increase the intensity of one's argument – but the unquestionably unethical ploy of using such tactics as a *substitute* for reasoning, as a "counterfeit" argument, in effect.

(I borrow this phrase from the title of Lionel Ruby's excellent book, *Fallacy: the Counterfeit of an Argument*.)

When you see this happening in a discussion, you can be fairly sure that the person resorting to such tactics is intellectually and morally bankrupt, and is more interested in winning than in seeking truth. That is your cue to find something better to do.

Chapter 9 – Thoughts on Induction and Infinity

Postscript 2 – arguments about infinity: Fred Seddon, in his review of this book (*Journal of Ayn Rand Studies*, December 2014), challenges my refutation of Cantor's argument that the number of counting numbers and even numbers in infinite sets are the same. I say that the ratio is always 2:1, not 1:1, for any finite set, however small or large, of such numbers, and there is no justification for saying that the ratio is any different when those sets are indefinitely large (infinite).

Sedden offers several dubious objections to this.

> (1) He claims that Cantor "proved" that there is a one-to-one correspondence between counting numbers and even numbers, when in fact the literal identity is the one-to-one correspondence between the even counting numbers and the even numbers; and as I pointed out, while these two groups of numbers are identically infinite, there is another group of numbers – the odd counting numbers – which is just as large and when combined with the even counting numbers makes the total set of counting numbers twice as large as the even numbers.

> (2) He claims that it is illicit to extend inductions based on finite groups of numbers to infinite groups. However, this argument cuts both ways. If it rules out the induction that counting-to-even is always 2:1, then it rules out the induction that one-to-one between counting and even is always extendable.

(3) He argues that since it's illicit to argue that something true for any x is also true for any non-x, and that since the infinite is non-finite, then you cannot argue that anything true of the finite is also true of the infinite. Again, if this is so, then neither can you argue that any conclusion about one-to-one for finite sets of countings and evens is true for infinite sets of countings and evens – nor, in particular, can you argue that just because you can *make* a one-to-one between finite sets of countings and evens, you can also make a one-to-one between *infinite* sets of countings and evens.

(4) He suggests that because "the privative makes infinite the contradictory of the finite," therefore it is a contradiction to argue that what is true of the finite must also be true of the infinite. Once

143

again, this would imply that from the fact that you can set up a one-to-one correspondence between the each of the finite counting and even numbers, you cannot infer that there is a one-to-one correspondence between infinite sets of those numbers.

Setting aside all of these technical variations on the same error, Cantor's argument can be more simply seen as equivalent to a claim that despite the fact that two parallel lines are always the same distance apart along any finite extent of those lines, if they were infinitely extended, they could meet, or become half or twice as far apart, or turn into a duck and a rabbit. Once you try to prove things about infinity by discarding the evidence of your senses and the common sense conclusions of your thinking, there is no limit to the counter-intuitive nonsense you can claim to be able to "prove."

Appendix: Studies on the Pythagorean Theorem

In the first section of this paper, I will derive the mathematical formulas I discovered for two infinite classes of right triangles with all sides of integral length (known as *Pythagorean triples*). Then I will generalize to all such classes. Finally I will identify the limits on the ratios of the sides to one another throughout all such classes.

Part I: Two infinite classes of right triangles with integral-length sides

The Pythagorean Theorem is a mathematical law about the relationship between the sides of a right triangle – i.e., a triangle with the two smaller sides perpendicular to each other, forming a right angle (90°). The Theorem, which dates back to ancient Greece, is a staple of the standard 10th grade geometry course, many children being introduced to it in elementary school.

Basically, the Theorem states that, for any right triangle, xyz (z being the longest side, the hypotenuse), the sum of the squares of the lengths of the two smaller sides is equal to the square of the length of the largest side: $x^2 + y^2 = z^2$.

For instance, if $x = 2$ and $y = 3$, then $z^2 = x^2 + y^2 = 4 + 9 = 13$. Since $z^2 = 13$, z is $13^{1/2}$, or approximately 3.6.

There are non-integral values which satisfy the Pythagorean equation. However, for ease of computation and observing numerical relationships, I chose to limit myself, at least initially, to considering triangles whose sides were all whole numbers.

Since I didn't know very many such triangles (the main one that came to mind being the ones with sides $x = 3$, $y = 4$, $z = 5$), this necessarily led to a lot of fishing around for triangles that would fit the Theorem with whole-number values for the sides. The payoff, of course, would be to find enough of them that some pattern common to them all might emerge.

This approach succeeded very well, and it yielded results I had never seen before and certainly did not expect.

One thing I *did* know about right triangles with whole-number lengths on each side was that the two smaller sides could not be equal. (Otherwise, the third side's length would have a square root in it.) In other words, the second side, y, would have to be some quantity, a, larger than side y, so that $y = x + a$.

To be precise, I have to say that y must be some quantity, a, *different from* side x, because for some solutions we will see later, y is actually smaller than x. We will continue to use the expression $x + a$ throughout, however, simply keeping in mind that sometimes the amount a will be a negative quantity that we add to x in order to get a smaller number, y.

(This is the exact equivalent of subtracting a positive amount, a, with the added advantage that we can keep the form $y = x + a$, for consistency during our explorations.)

As for the third side, the largest side of the triangle, it would naturally be some quantity, b, larger than side x, such that: $z = x + b$. (Since side z is larger than side y, amount b will always be larger than amount a.

Also, unlike amount a, amount b will always be a positive number, since side z will always be larger than side x, whereas side y sometimes turns out not to be larger than side x.)

Using these two equivalent expressions for y and z, we can reformulate the Pythagorean Theorem equation, $x^2 + y^2 = z^2$, as: $x^2 + (x + a)^2 = (x + b)^2$.

If we then multiply out the terms of this equation and combine like terms, we get a second useful form of the equation: $x^2 + 2ax - 2bx + a^2 - b^2 = 0$. During our explorations, we will frequently make use of one or the other of these forms of the equation.

Now, what do we find, when we proceed empirically, step by step, by trial-and-error? Here is the methodology I followed:

Let's begin by setting amount a equal to the smallest quantity that will still yield a whole number when added to x – namely, the number 1.

Then, let's explore the solutions that exist for x, when we let b taken on all the whole-number values it can take that allow $x + b$ to be a whole number greater than x or $x + a$ (namely, the numbers 2, 3, 4, 5, etc). (We will ignore negative solutions for x, because they will cause $x + a$ to be 0, thus collapsing the Pythagorean triangle's second side.)

Finally, let's see what relationship there may be between a and b or between a, b, and x, when a particular value for b yields a whole-number solution for x. And let's try to observe, if possible, how the other values for b fail to yield such a solution.

After this, we will do the same thing for $a = 2$, letting b range through 3, 4, 5, 6, etc. – then for $a = 3$, with b ranging through 4, 5, 6, 7, etc. – then for $a = 4$, letting b range through 5, 6, 7, 8, etc. By then, a pattern should emerge.

$\underline{a = 1, b = 2}$:

$x^2 + 2(1)x = 2(2)x + 1^2 - 2^2 = 0$

$x^2 + 2x - 4x + 1 - 4 = 0$

$x^2 - 2x - 3 = 0$

Factoring and solving for x:

$(x - 3)(x + 1) = 0$

$x = 3$ (ignoring the negative solution, $x = -1$)

$x + a = 4, x + b = 5$

$3^2 + 4^2 = 9 + 16 = 25 = ? = 5^2$

Yes, this is a solution.

Omitting (for the sake of space) the first two steps in deriving each of the following, we find that for:

$b = 3$, $x^2 - 4x - 8 = 0$, not factorable, no whole-number solution.

$b = 4$, $x^2 - 6x - 15 = 0$, not factorable, no whole-number solution.

$b = 5$, $x^2 - 8x - 24 = 0$, not factorable, no whole-number solution.

$b = 6$, $x^2 - 10x - 35 = 0$, not factorable, no whole-number solution.

$b = 7$, $x^2 - 12x - 48 = 0$, not factorable, no whole-number solution.

$b = 8$, $x^2 - 14x - 63 = 0$, not factorable, no whole-number solution.

$b = 9$, $x^2 - 16x - 80 = 0$

$(x - 20)(x + 4) = 0$

$x = 20$ (ignoring the negative solution, $x = -4$)

$x + a = 21$, $x + b = 29$

$20^2 + 21^2 = 400 + 441 = 841 = ? = 29^2$

Yes, this is a solution.

So, to this point, we have whole-number solutions when $a = 1$, $b = 2$, $x = 3$ and when $a = 1$, $b = 9$, $x = 20$. We could continue indefinitely with increasing values of b, and we could find other whole-number solutions for x.

However, we will stop here and move on to other values for a, because we now have enough results (when combined with results of using other values of a) in order to see the patterns that will emerge.

We will now proceed, as above, with $a = 2$ (letting b range through 3, 4, 5, 6, etc.). Rather than continuing to use the quadratic form of the Pythagorean Theorem equation, however, let's solve it for x and use it in that form.

$x^2 + (x + a)^2 = (x + b)^2$

$x^2 + x^2 + 2ax + a^2 = x^2 + 2bx + b^2$

$x^2 + 2ax - 2bx + a^2 - b^2 = 0$

$x^2 + 2x(a - b) + (a^2 - b^2) = 0$

$x^2 + 2x(a - b) + (a - b)^2 - (a - b)^2 + (a^2 - b^2) = 0$

$[x + (a - b)]^2 - (a - b)^2 + (a^2 - b^2) = 0$

$[x + (a - b)]^2 - a^2 + 2ab - b^2 + a^2 - b^2 = 0$

$[x + (a - b)]^2 + 2b(a - b) = 0$

$[x + (a - b)]^2 - 2b(b - a) = 0$

$[x + (a - b)]^2 = 2b(b - a)$

$x + (a - b) = [2b(b - a)]^{1/2}$

$x = [2b(b - a)]^{1/2} - (a - b)$

$x = [2b(b - a)]^{1/2} + (b - a)$

Now let's proceed, with $a = 2$, allowing b to range over 3, 4, 5, 6, etc.:

$b = 3$, $x = [6(1)]^{1/2} + 1 = 6^{1/2} + 1$, no whole-number solution.

$b = 4$, $x = [8(2)]^{1/2} + 2 = 16^{1/2} + 2 = 4 + 2 = 6$.

$x + a = 8$, $x + b = 10$

$6^2 + 8^2 = 36 + 64 = 100 = ? + 10^2$

Yes, this is a solution.

$b = 5$, $x = [10(3)]^{1/2} + 3 = 30^{1/2} + 3$, no whole-number solution.

$b = 6$, $x = [12(4)]^{1/2} + 4 = 4(3)^{1/2} + 4$, ditto.

$b = 7$, $x = [14(5)]^{1/2} + 5 = 70^{1/2} + 5$, ditto.

$b = 8$, $x = [16(6)]^{1/2} + 6 = 4(6)^{1/2} + 6$, ditto.

$b = 9$, $x = [18(7)]^{1/2} + 7 = 3(14)^{1/2} + 7$, ditto.

$b = 10$, $x = [20(8)]^{1/2} + 8 = 4(10)^{1/2} + 8$, ditto.

$b = 11$, $x = [22(9)]^{1/2} + 9 = 3(22)^{1/2} + 9$, ditto.

$b = 12$, $x = [24(10)]^{1/2} + 10 = 4(15)^{1/2} + 10$, ditto.

$b = 13$, $x = [26(11)]^{1/2} + 11 = 286^{1/2} + 11$, ditto.

$b = 14$, $x = [28(12)]^{1/2} + 12 = 4(21)^{1/2} + 12$, ditto.

$b = 15$, $x = [30(13)]^{1/2} + 13 = 390^{1/2} + 13$, ditto.

$b = 16$, $x = [32(14)]^{1/2} + 14 = 8(7)^{1/2} + 14$, ditto.

$b = 17$, $x = [34(15)]^{1/2} + 15 = 510^{1/2} + 15$, ditto.

$b = 18$, $x = [36(16)]^{1/2} + 16 = 576^{1/2} + 16 = 24 + 16 = 40$.

$x + a = 42$, $x + b = 58$

$40^2 + 42^2 = 1600 + 1764 = 3364 = ? = 58^2$

Yes, this is a solution.

To this point, we have whole-number solutions for x when $a = 1$, $b = 1$, $x = 3$; when $a = 1$, $b = 9$, $x = 20$; when $a = 2$, $b = 4$, $x = 6$; and when $a = 2$, $b = 18$, $x = 40$.

Again, we could continue looking for more whole-number solutions of x; but, as before, it's cumbersome and not necessary. Instead, let's put the solutions we do have into a table and see what relationships may be emerging:

a	b	x	x+a	x+b	a/b	a+b	2(a+b)	x
1	2	3	4	5	1/2	3		
2	4	6	8	10	1/2	6		
1	9	20	21	29	1/9		20	20
2	18	40	42	58	1/9		40	40

In the right-hand section of the table (i.e., the last four columns), I have shown the relationships discovered in these solutions. For $a = 1$, $b = 2$ and for $a = 2$, $b = 4$, it is true that $a/b = 1/2$, or $a = b/2$, and that $x = a + b$. For $a = 1$, $b = 9$ and for $a = 2$, $b = 18$, it is true that $a = b/9$, and that $x = 2(a + b)$.

We now have two distinct groups of whole-number solutions which are multiples of $a = 1$, $b = 2$ and of $a = 1$, $b = 9$. This suggests that there are additional solutions in each group. Let's demonstrate a couple of examples and then prove the relationships derived above to be valid as general formulas.

For $x = a + b$, $a = b/2$:

If $a = 3$, then $b = 6$ and $x = 9$.

So $x + a = 12$, $x + b = 15$.

$9^2 + 12^2 = 81 + 144 = 225 = ? = 15^2$ (yes)

If $a = 4$, then $b = 8$ and $x = 12$.

So $x + a = 16$, $x + b = 20$.

$12^2 + 16^2 = 144 + 256 = 400 = ? + 20^2$ (yes)

Appendix: Studies on the Pythagorean Theorem

For $x = 2a + b$, $a = b/9$:

If $a = 3$, then $b = 27$ and $x = 60$.

So $x + a = 63$, $x + b = 87$.

$60^2 + 63^2 = 3600 + 3969 = 7569 = ? = 87^2$ (yes)

If $a = 4$, then $b = 36$ and $x = 80$.

So $x + a = 84$, $x + b = 116$.

$80^2 + 84^2 = 6400 + 7056 = 13{,}546 = ? = 116^2$ (yes)

With these new results, we can expand the previous table as follows:

a	b	x	x+a	x+b	a/b	a+b	2(a+b)	x
1	2	3	4	5	1/2	3		
2	4	6	8	10	1/2	6		
3	6	9	12	15	1/2	9		
4	8	12	16	20	1/2	12		
1	9	20	21	29	1/9		20	20
2	18	40	42	58	1/9		40	40
3	27	60	63	87	1/9		60	60
4	36	80	84	116	1/9		80	80

After seeing the patterns emerging so vividly in the preceding charts, it may seem somewhat unnecessary or anticlimactic to *logically prove* that the relations always hold. After all, haven't we already *seen*, empirically, that they must?

Nonetheless, it is important to "see" it *rationally* as well, for it illustrates that there are often not just one, but two paths for discovering new facts or principles: the *inductive* (as above) and the *deductive*.

Substitute $x = a + b$ into $x^2 + (x + a)^2 = (x + b)^2$, and solve for a in terms of b:

$$(a + b)^2 + (a + b + a)^2 = (a + b + b)^2$$

$$(a + b)^2 + (2b + a)^2 = (a + 2b)^2$$
$$a^2 + 2ab + b^2 + 4a^2 + 4ab + b^2 = a^2 + 4ab + 4b^2$$
$$4a^2 + 2ab - 2b^2 = 2a^2 + ab - b^2 = 0$$
$$(2a - b)(a + b) = 0$$
$a = b/2, -b$ (ignore latter solution)

Check by substituting $a = b/2$ into $x^2 + (x + a)^2 = (x + b)^2$ and solving for x:

$$x^2 + (x + b/2)^2 = (x + b)^2$$
$$x^2 + x^2 + bx + b^2/4 = x^2 + 2bx + b^2$$
$$x^2 - bx - 3b^2/4 = 0$$
$$4x^2 - 4bx - 3b^2 = 0$$
$$(2x - 3b)(2x + b) = 0$$
$x = 3b/2, -b/2$ (ignore latter solution)
$x = 3b/2 = (1 + 2)b/2 = b/2 + b = a + b$
$x = a + b$ (This completes the check)

Substitute $x = 2a + b$ into $x^2 + (x + a)^2 = (x + b)^2$, and solve for a in terms of b:

$$[2(a + b)]^2 + [2(a + b) + a]^2 = [2(a + b) + b]^2$$
$$4a^2 + 8ab + 4b^2 + 9a^2 + 6ab + 4b^2 = 4a^2 + 6ab + 9b^2$$
$$9a^2 + 8ab - b^2 = 0$$
$$(9a - b)(a + b) = 0$$
$a = b/9, -b$ (ignore latter solution)

Check by substituting $a = b/9$ into $x^2 + (x + a)^2 = (x + b)^2$, and solving for x:

$$x^2 + (x + b/9)^2 = (x + b)^2$$
$$x^2 + x^2 + 2bx/9 + b^2/81 = x^2 + 2bx + b^2$$
$$x^2 - 16bx/9 - 80b^2/81 = 0$$

$(x + 4b/9)(x - 20b/9) = 0$

$x = 20b/9, -4b/9$ (ignore latter solution)

$x = 20b/9 = (2 + 18)b/9 = 2b/9 + 2b = 2(b/9 + b)$

$x = 2(a + b)$ (This completes the check.)

[We will note in passing that there are alternate solutions for x which are of some interest but, presently, of uncertain usefulness:

$x = 3b/2 = 3b/4 + 3b/4 = 3a/2 + 3b/4 = 3(a + b/2)/2$

$x = 20b/9 = 10b/9 + 10b/9 = 10a + 10b/9 = 10(a + b/9)$]

Part II: Generalizing to an infinite number of infinite classes of right triangles with integral-length sides

We have just seen both inductively and deductively that two infinite classes of right triangles with integral-length sides exist – specifically, when $a = b/2$ and $x = a + b$, and when $a = b/9$ and $x = 2(a + b)$. This suggests the possibility that there are additional such infinite classes for higher integral factors of $(a + b)$.

We will now demonstrate that, in fact, there is an *infinite* number of such classes where $x = c(a + b)$, for any positive integer, c. We will also derive the general expression of the relationship between a and b in terms of c.

Let's begin by substituting a series of integral multiples of $(a + b)$ for x in the equation $x^2 + 2ax - 2bx + a^2 - b^2 = 0$.

Let $x = 3(a + b)$:

$[3(a + b)]^2 + 2a[3(a + b)] - 2b[3(a + b)] + a^2 - b^2 = 0$

$9a^2 + 18ab + 9b^2 + 6a^2 + 6ab - 6ab - 6b^2 + a^2 - b^2 = 0$

$16a^2 + 18ab + 2b^2 = 0$

$(16a + 2b)(a + b) = 0$

$a = -2b/16 = -b/8$ (ignore latter solution)

If $a = -1$, then $b = 8$, $x = 21$, $x + a = 20$, $x + b = 29$

$21^2 + 20^2 = 241 + 200 = 441 = ? = 29^2$ Yes!

[Note: this result is a flip-flop of x and $x + a$ from the result of $x = 2(a + b)$. Now x is *larger* than $x + a$, so a is actually a *negative* quantity in this case. This is a significant development, which holds true for larger multiples of $(a + b)$, as well. We will identify the reason for this flip-flop of Part IV of this essay.]

Let $x = 4(a + b)$:

$16a^2 + 32ab + 16b^2 + 8a^2 + 8ab - 8ab - 8b^2 + a^2 - b^2 = 0$

$25a^2 + 32ab + 7b^2 = 0$

$(25a + 7b)(a + b) = 0$

$a = -7b/25$ (ignore other solution)

If $a = -7$, then $b = 25$, $x = 72$, $x + a = 65$, $x + b = 97$

$72^2 + 65^2 = 5184 + 4225 = 9409 = ? = 97^2$ Yes!

Let $x = 5(a + b)$:

$25a^2 + 50ab + 25b^2 + 10a^2 + 10ab - 10ab - 10b^2 + a^2 - b^2 = 0$

$36a^2 + 50ab + 14b^2 = 0$

$(36a + 14b)(a + b) = 0$

$a = -14b/36 = -7b/18$ (ignore other solution)

If $a = -7$, then $b = 18$, $x = 55$, $x + a = 48$, $x + b = 73$

$55^2 + 48^2 = 3025 + 2305 = 5329 = ? = 73^2$ Yes!

Let $x = 6(a + b)$:

$36a^2 + 72ab + 36b^2 + 12a^2 + 12ab - 12ab - 12b^2 + a^2 - b^2 = 0$

$49a^2 + 72ab + 23b^2 = 0$

$(49a + 23b)(a + b) = 0$

$a = -23b/49$ (ignore other solution)

If $a = -23$, then $b = 49$, $x = 156$, $x + a = 133$, $x + b = 205$

$156^2 + 133^2 = 24{,}336 + 17{,}689 = 42{,}025 = ? = 205^2$ Yes!

Let $x = 7(a + b)$:

$49a^2 + 98ab + 49b^2 + 14a^2 + 14ab - 14ab - 14b^2 + a^2 - b^2 = 0$

$64a^2 + 98ab + 34b^2 = 0$

$(64a + 34b)(a + b) = 0$

$a = -34b/64 = -17b/32$ (ignore other solution)

If $a = -17$, then $b = 32$, $x = 105$, $x + a = 88$, $x + b = 137$

$105^2 + 88^2 = 11{,}025 + 7744 = 18{,}769 = ? = 137^2$ Yes!

Let $x = 8(a + b)$:

$64a^2 + 128ab + 64b^2 + 16a^2 + 16ab - 16ab - 16b^2 + a^2 - b^2 = 0$

$81a^2 + 128ab + 47b^2 = 0$

$(81a + 47b)(a + b) = 0$

$a = -47/81$ (ignore other solution)

If $a = -47$, then $b = 81$, $x = 272$, $x + a = 225$, $x + b = 353$

$272^2 + 225^2 = 73{,}984 + 50{,}625 = 124{,}609 = ? = 353^2$ Yes!

Let $x = 9(a + b)$:

$81a^2 + 162ab + 81b^2 + 18a^2 + 18ab - 18ab - 18b^2 + a^2 - b^2 = 0$

$100a^2 + 162ab + 62b^2 = 0$

$(100a + 62b)(a + b) = 0$

$a = -62b/100 = -31b/50$ (ignore other solution)

If $a = -31$, then $b = 50$, $x = 171$, $x + a = 140$, $x + b = 221$

$171^2 + 140^2 = 29{,}241 + 19{,}600 = 48{,}841 = ? = 221^2$ Yes!

Let $x = 10(a + b)$:

$100a^2 + 200ab + 100b^2 + 20a^2 + 20ab - 20ab - 20b^2 + a^2 - b^2 = 0$

$121a^2 + 200ab + 79b^2 = 0$

$(121a + 79b)(a + b) = 0$

$a = -79b/121$ (ignore other solution)

If $a = -79$, then $b = 121$, $x = 420$, $x + a = 341$, $x + b = 541$

$420^2 + 341^2 = 176{,}400 + 116{,}281 = 292{,}681 = ? = 541^2$ Yes!

So far, we have found whole-number solutions for the Pythagorean equation, as follows:

$x = (a+b)$, $a = b/2$
$x = 2(a+b)$, $a = b/9$
$x = 3(a+b)$, $a = -b/8$
$x = 4(a+b)$, $a = -7b/25$
$x = 5(a+b)$, $a = -7b/18$
$x = 6(a+b)$, $a = -23b/49$
$x = 7(a+b)$, $a = -17b/32$
$x = 8(a+b)$, $a = -47b/81$
$x = 9(a+b)$, $a = -31b/50$
$x = 10(a+b)$, $a = -79b/121$

We note that when the constant factor of $(a + b)$ is 2, 4, 6, 8, or 10, the denominator of the fraction is a perfect square (9, 25, 49, 81, 121). By taking the other fractions – where the constant factor of $(a + b)$ is 1, 3, 5, 7, or 9 – *out of* lowest terms, we can put their denominators into perfect squares, too, as follows:

$x = 1(a+b)$, $a = b/2 = 2b/4 = 2b/2^2$
$x = 2(a+b)$, $a = b/9 = = b/3^2$
$x = 3(a+b)$, $a = -b/8 = -2b/16 = -2b/4^2$
$x = 4(a+b)$, $a = -7b/25 = = -7b/5^2$
$x = 5(a+b)$, $a = -7b/18 = -14b/36 = -14b/6^2$
$x = 6(a+b)$, $a = -23b/49 = = -23b/7^2$
$x = 7(a+b)$, $a = -17b/32 = -34b/64 = -34b/8^2$
$x = 8(a+b)$, $a = -47b/81 = = -47b/9^2$
$x = 9(a+b)$, $a = -31b/50 = -62b/100 = -62b/10^2$

Appendix: Studies on the Pythagorean Theorem

$$x = 10(a + b), \ a = -79b/121 \qquad = -79b/11^2$$

Now we can see a clear pattern emerging. If we refer to the constant factor of $(a + b)$ as c, we can see that the denominator of the fraction (to which a is equal) for a given c is equal to the square of the next higher number (i.e., to $c + 1$).

For instance, the denominator of $b/9$, where $c = 2$, is equal to $(c + 1)^2 = (2 + 1)^2 = 3^2 = 9$. And the denominator of $-7b/25$, where $c = 4$, is equal to $(c + 1)^2 = (4 + 1)^2 = 5^2 = 25$. And so on.

If we next look at the numerators of the fraction (to which a is equal), we can note another pattern, less obvious than the one just observed for the denominator, but no less clear and real. The factors of b in the numerator (let us call them NFb) progress as follows:

$(c-1)^2$	c	NFb
0	1	$2 = 2 - 0 = 2 - 0^2 = 2 - (c-1)^2$
1	2	$1 = 2 - 1 = 2 - 1^2 = 2 - (c-1)^2$
4	3	$-2 = 2 - 4 = 2 - 2^2 = 2 - (c-1)^2$
9	4	$-7 = 2 - 9 = 2 - 3^2 = 2 - (c-1)^2$
16	5	$-14 = 2 - 16 = 2 - 4^2 = 2 - (c-1)^2$
25	6	$-23 = 2 - 25 = 2 - 5^2 = 2 - (c-1)^2$

Disregarding the minus sign, we can see that NFb is always two (2) smaller for any given number, c, than the square of the next lower number, $c - 1$. More precisely, for a given number, c, the factor of b in the numerator is equal to $2 - (c - 1)^2$, where $(c - 1)^2$ is the square of the next smaller number, $c - 1$.

So, $2 - (c - 1)^2$ is the factor of b in the numerator of the fraction to which a is equal, for any given whole number, c, in the equation $x = c(a + b)$. Combining the results for the numerator and denominator, we now see that when $x = c(a + b)$, $a = [2 - (c - 1)^2]b/(c + 1)2$.

Another way of arriving at the expression for the numerator factor, NFb, is as follows:

c	a	NFb
1	$2b/4$	2
2	$1b/9$	1
3	$-2b/16$	-2
4	$-7b/25$	-7
5	$-14b/36$	-14
6	$-23b/49$	-23

For a given number, c, we can see that when, from c, is subtracted the NFb for the next smaller number, $c - 1$, the remainder is equal to the product of the next smaller number with the second smaller number – i.e., to $(c - 1)(c - 2)$. For example, when $c = 5$, $5 - (-7) = 4(3)$. And when $c = 6$, $6 - (-14) = 5(4)$.

Since this relationship holds for a given number, c, we can also see that for the next higher number, $c + 1$, it is true that $(c + 1) - NFb_c = c(c - 1)$. Therefore, $NFb_c = (c + 1) - c(c - 1) = c + 1 - c^2 + c = -c^2 + 2c + 1 = -(c^2 - 2c - 1) = -(c^2 - 2c - 1 + 2 - 2) = -(c^2 - 2c + 1 - 2) = 2 - (c^2 - 2c + 1) = 2 - (c - 1)^2$, as above.

A third way of arriving at the expression for NFb is as follows:

c	a	NFb		$2 - NFb$		
1	$2b/4$	2		$2 -$	2	$= 0$
2	$1b/9$	1		$2 -$	1	$= 1$
3	$-2b/16$	$-2 =$	$1 - 3 = 2 - 4$	$2 - (-2)$		$= 4$
4	$-7b/25$	$-7 =$	$-2 - 5 = 2 - 9$	$2 - (-7)$		$= 5$
5	$-14b/36$	$-14 =$	$-7 - 7 = 2 - 16$	$2 - (-4)$		$= 16$
6	$-23b/49$	$-23 =$	$-14 - 9 = 2 - 25$	$2 - (-23)$		$= 25$

Note that NFb is:

    1 less when $c = 2$, than for $c = 1$

    3 less when $c = 3$, than for $c = 2$

4 less when $c = 3$, than for $c = 1$

5 less when $c = 4$, than for $c = 3$

9 less when $c = 4$, than for $c = 1$

7 less when $c = 5$, than for $c = 4$

16 less when $c = 5$, than for $c = 1$

9 less when $c = 6$, than for $c = 5$

25 less when $c = 6$, than for $c = 1$

Again, we see perfect squares – in this case, in the difference between the value for NFb, when $c = 1$ (namely, 2) and the numerator factor, NFb, for any given value for c. And the square in each case turns out to be the value of the square of the next smaller number, $c - 1$. For instance, when $c = 3$, $2 - (-2) = (3 - 1)^2 = 2^2 = 4$. When $c = 5$, $2 - (-14) = (5 - 1)^2 = 4^2 = 16$.

So, for a given whole number, c, the numerator factor added to the square of the next smaller whole number, $c - 1$, is always equal to 2. Or, the numerator factor, NFb, for a given whole number, c, is always $2 - (c - 1)^2$.

(Also, in passing, let us observe that if: $a = [2 - (c - 1)^2]b/(c + 1)^2$, then $a/b = [2 - (c - 1)^2]/(c + 1)^2$. These results will have some usefulness in the following section of this essay, as well as in final section, which deals with the limits of the ratios of the sides to one another, for all the infinite classes of integral-length-sided right triangles.)

Part III: Extending the generalization even further

So far, we have worked only with positive integral factors in $x = c(a + b)$. Now we will show that, while c need not be a positive integer, it must be a rational number that falls within a certain domain.

First, we will consider whether there are fractional multiples of $(a + b)$ that satisfy $x = c(a + b)$ and $a = [2 - (c - 1)^2]b/(c + 1)^2$. Then we will consider whether *every* possible fractional multiple of $(a + b)$ – i.e., the product of every rational number, n, and $(a + b)$ satisfies $x = c(a + b)$ and $a = [2 - (c - 1)^2]b/(c + 1)^2$ – and, thirdly, whether *only* (whether some or all) rational numbers do so.

How the Martians Discovered Algebra

First, a few helpful examples that satisfy our criterion:

Let $x = (a + b)/2$, $c = 1/2$

$(a + b)^2/4 + [(a + b)/2 + a]^2 = [(a + b)/2 + b]^2$

$(a + b)^2 + (a + b)^2 + 4a(a + b) + 4a^2 = (a + b)^2 + 4b(a + b) + 4b^2$

$a^2 + 2ab + b^2 + 4a^2 + 4ab + 4a^2 = 4ab + 4b^2 + 4b^2$

$9a^2 + 2ab - 7b^2 = 0$

$(9a - 7b)(a + b) = 0$

$a = 7b/9$ (ignoring $a = -b$)

If $a = 7$, $b = 9$, then $x = 8$, $x + a = 15$, $x + b = 17$

Check: $8^2 + 15^2 = 64 + 225 = 289 = 17^2$

Check: $7 = [2 - (c - 1)^2](9)/(c + 1)^2$

$7c^2 + 14c + 7 = 18 - 9c^2 + 18c - 9$

$16c^2 - 4c - 2 = 0$

$(4c - 2)(4c + 1) = 0$

$c = 1/2, -1/4$

Let $x = (a + b)/4$, $c = 1/4$. Omitting the first three steps from here on, for brevity:

$25a^2 + 2ab - 23b^2 = 0$

$(25a - 23b)(a + b) = 0$

$a = 23b/25$ (ignoring $a = -b$)

If $a = 23$, $b = 25$, then $x = 12$, $x + a = 35$, $x + b = 37$

Check: $12^2 + 35^2 = 144 + 1225 = 1369 = 37^2$

Check: $23 = [2 - (c - 1)^2](25)/(c + 1)^2$

$23c^2 + 46c + 23 = 50 - 25c^2 + 50c - 25$

$48c^2 - 4c - 2 = 0$

$(4c - 1)(12c + 2) = 0$

$c = 1/4, -1/6$

Let $x = 3(a + b)/4$, $c = 3/4$.

$49a^2 + 18ab - 31b^2 = 0$

$(49a - 31b)(a + b) = 0$

$a = 31b/49$ (ignoring $a = -b$)

If $a = 31$, $b = 49$, then $x = 60$, $x + a = 91$, $x + b = 109$

Check: $60^2 + 91^2 = 3600 + 8281 = 11{,}881 = 109^2$

Check: $31c^2 + 62c + 31 = 98 - 49c^2 + 98c - 49$

$80c^2 - 36c - 18 = 40c^2 - 18c - 9 = 0$

$(4c - 3)(10c + 3) = 0$

$c = 3/4, -3/10$

Let $x = 5(a + b)/4$, $c = 5/4$.

$81a^2 + 50ab - 31b^2 = 0$

$(81a - 31b)(a + b) = 0$

$a = 31b/81$ (ignoring $a = -b$)

If $a = 31$, $b = 81$, then $x = 140$, $x + a = 171$, $x + b = 221$

Check: $140^2 + 171^2 = 19{,}600 + 29{,}241 = 48{,}841 = 221^2$

Check: $31c^2 + 62c + 31 = 162 - 81c^2 + 162c - 81$

$112c^2 - 100c - 50 = 56c^2 - 50c - 25 = 0$

$(4c - 5)(14c + 5) = 0$

$c = 5/4, -5/14$

Now let's see if there is a *general* formula that fits cases where n/d, the factor of $(a + b)$ ranges from a common fraction to an integer to a mixed fraction. Here are some examples with a constant denominator, increasing numerator:

$x = (1/4)(a + b)$, $a = 23b/25 = [(1 + 4)^2 - 2(1^2)]b/(1 + 4)^2$

$x = (2/4)(a + b)$, $a = 7b/9 = 28b/36 = [(2 + 4)^2 - 2(2^2)]b/(2 + 4)^2$

$x = (3/4)(a + b)$, $a = 31b/49 = [(3 + 4)^2 - 2(3^2)]b/(3 + 4)^2$

$x = (4/4)(a+b)$, $a = 1b/2 = 32b/64 = [(4+4)^2 - 2(4^2)]b/(4+4)^2$

$x = (5/4)(a+b)$, $a = 31b/81 = [(5+4)^2 - 2(5^2)]b/(5+4)^2$

$x = (n/4)(a+b)$, $a = [(n+4)^2 - 2(n^2)]b/(n+4)^2$

Here are some examples where the denominator and numerator both increase:

$x = (1/2)(a+b)$, $a = 7b/9 = [(1+2)^2 - 2(1^2)]b/(1+2)^2$

$x = (3/3)(a+b)$, $a = b/2 = 18b/36 = [(3+3)^2 - 2(3^2)]b/(3+3)^2$

$x = (5/4)(a+b)$, $a = 31b/81 = [(5+4)^2 - 2(5^2)]b/(5+4)^2$

$x = (n/d)(a+b)$, $a = [(n+d)^2 - 2(n^2)]b/(n+d)^2$

It seems, then, that a reasonable hypothesis for *all* positive rational numbers, n/d, where $x = (n/d)(a+b)$, is:

$$a = [(n+d)^2 - 2(n^2)]b/(n+d)^2$$
$$= (n^2 + 2nd + d^2 - 2n^2)b/(n^2 + 2nd + d^2)$$
$$= (d^2 + 2nd - n^2)b/(n^2 + 2nd + d^2).$$

Let us deductively prove this hypothesis. If $x = n(a+b)/d$, then:

$$(n/d)^2(a+b)^2 + [(n/d)(a+b) + a]^2 = [(n/d)(a+b) + b]^2$$
$$(n/d)^2(a+b)^2 + [(n/d)(a+b) + a]^2 = [(n/d)(a+b) + b]^2$$
$$n^2(a^2 + 2ab + b^2) + d^2([n^2(a+b)^2/d^2] + [2an(a+b)/d] + a^2)$$
$$= d^2([n^2(a+b)/d^2] + [2bn(a+b)/d] + b^2)$$
$$a^2n^2 + 2abn^2 + b^2n^2 + 2a^2nd + a^2d^2 + 2b^2nd + b^2d^2 = 0$$
$$a^2(n^2\ 2nd + d^2) + 2abn^2 - b^2(d^2 + 2nd - n^2) = 0$$
$$[a(n^2 + 2nd + d^2) - b(d^2 + 2nd - n^2)][a+b] = 0$$
$$a = (d^2 + 2nd - n^2)b/(n^2 + 2nd + d^2).\ \text{Q.E.D.!}$$

Now, can we find a simpler form for a that be used? If $c = n/d$, where c is any positive rational number, n/d (the ratio of any positive integers, n and d), then:

$$a = (d^2 + 2nd - n^2)b/(n^2 + 2nd + d^2)$$
$$= (d^2 + 2nd - n^2)b/d^2/(n^2 + 2nd + d^2)/d^2$$
$$= [1 + (2n/d) - n^2/d^2]b/[n^2/d^2 + (2n/d) + 1]$$
$$= (1 + 2c - c^2)b/(c^2 + 2c + 1).$$

So, yes, this simpler form, $a = (1 + 2c - c^2)b/(c^2 + 2c + 1)$, can be used for any positive rational number, c. Here is an example using both formulas:

$$x = 2(a + b)/3, c = 2/3, n = 2, d = 3$$
$$a = (1 + 2c - c^2)b/(c^2 + 2c + 1)$$
$$= [1 + (4/3) - (4/9)]b/[(4/9) + (4/3) + 1]$$
$$= 17b/9/25/9$$
$$= 17b/25$$

Are there any *negative* rational values of c, for which this relationship holds? As it turns out, yes, the relationship holds for any negative rational value of c, where $c < -1$. However, when $c = -1$, a/b is undefined or infinite in value: $a/b = (1 - 2 - 1)/(1 - 2 + 1) = -2/0$.

It also so happens that there are no triples defined for values of c between 0 and -1, and that the triples defined by negative rational values of c smaller than -1 are "flip-flopped" versions $(x > x + a)$ of the triples defined by values of c between 0 and 1, because the corresponding values for a are all negative. Here are a few representative examples of negative rational values for c that, respectively, do and do not yield positive integral Pythagorean triples:

$c = -2, x = 12, x + a = 5, x + b = 13.$

$c = -3, x = 15, x + a = 8, x + b = 17.$

$c = -4, x = 56, x + a = 33, x + b = 65.$

$c = -3/2, x = 24, x + a = 7, x + b = 25.$

$c = -5/3, x = 35, x + a = 12, x + b = 37.$

$c = -9/7, x = 63, x + a = 16, x + b = 65.$

$c = -4/3, x = 40, x + a = 9, x + b = 41.$

$c = -5/4$, $x = 60$, $x + a = 11$, $x + b = 61$.

$c = -1/2$, $a/b = -1/1$, $x = 0$, $x + a = -1$.

$c = -2/5$, $a/b = 1/19$, $x = -4$, $x + a = -3$.

$c = -1/3$, $a/b = 1/2$, $x = -1$, $x + a = 0$.

$c = -1/4$, $a/b = 7/9$, $x = -4$, $x + a = 3$.

$c = -2/3$, $a/b = -7/1$, $x = 4$, $x + a = -3$.

$c = -3/4$, $a/b = -17/1$, $x = 12$, $x + a = -5$.

Without showing all the details, there are positive integral solutions $x + b > x > x + a > 0$, when $c < -1$. There are *no* triples when $-1 \leq c \leq 0$, because:

1) when $c = -1$, x is infinite
2) when $-1 < c < -1/2$, $x > 0$, but $x + a < 0$
3) when $c = -1/2$, $x = 0$ and $x + a < 0$
4) when $-1/2 < c < -1/3$, $-1 < x < 0$ and $x + a < 0$
5) when $c = -1/3$, $x = -1$ and $x + a = 0$
6) when $-1/3 < c < 0$, $x < -1$ and $x + a = 0$
7) when $c = 0$, $x = 0$

Therefore, there are definable triples for values of c, such that $c < -1$, $0 < c < (2^{1/2} + 1)$, and $(2^{1/2} + 1) < c$. Also, the triples defined by $c < -1$ are flip-flopped from those defined by $0 < c < 1$, and the triples defined by $c = 1$ and $1 < c < (2^{1/2} + 1)$ are flip-flopped from those defined by $(2^{1/2} + 1) < c$. (As I will show in Part IV, the latter flip-flop point is defined by $c = (2^{1/2} + 1)$.)

We have yet to establish whether this relationship holds for only the positive rational numbers and the negative rational numbers less than -1, or whether there are some *irrational* real numbers for which it holds, as well. Let's first examine a couple of noteworthy examples where it does *not* hold.

First, let's use π. Suppose $x = \pi(a + b)$, $c = \pi$. Plugging into $x^2 + (x + a)^2 = (x + b)^2$, we get:

Appendix: Studies on the Pythagorean Theorem

$(\pi^2 a^2 + 2\pi^2 ab + \pi^2 b^2) + [\pi(a+b) + a]^2 = [\pi(a+b) + b]^2$

$\pi^2 a^2 + 2\pi^2 ab + \pi^2 b^2 + \pi^2(a+b)^2 + 2\pi(a+b) + a^2 = \pi^2(a+b)^2 + 2\pi b(a+b) + b^2$

$\pi^2 a^2 + 2\pi^2 ab + \pi^2 b^2 + 2\pi a^2 + a^2 - 2\pi b^2 - b^2 = 0$

$a^2(\pi^2 + 2\pi + 1) + 2\pi^2 ab + b^2(\pi^2 - 2\pi - 1) = 0$

$[a(\pi^2 + 2\pi + 1) - b(1 + 2\pi - \pi^2)](a+b) = 0$

$a = (1 + 2\pi - \pi^2) b/(\pi^2 + 2\pi + 1)$.

Now, when $b = \pi^2 + 2\pi + 1$, then $a = 1 + 2\pi - \pi^2$, so:

$x = \pi(a+b) = \pi(1 + 2\pi - \pi^2 + \pi^2 + 2\pi + 1) = \pi(4\pi + 2) = 4\pi^2 + 2\pi$

$x + a = 4\pi^2 + 2\pi + 1 + 2\pi - \pi^2 = 3\pi^2 + 4\pi + 1$

$x + b = 4\pi^2 + 2\pi + \pi^2 + 2\pi + 1 = 5\pi^2 + 4\pi + 1$

These are not integral values, but let's check by plugging these values for x, $x + a$, and $x + b$ into the Pythagorean equation:

$(4\pi^2 + 2\pi)^2 + (3\pi^2 + 4\pi + 1)^2$

$= 16\pi^4 + 16\pi^3 + 4\pi^2 + 9\pi^4 + 24\pi^3 + 22\pi^2 + 8\pi + 1$

$= 25\pi^4 + 40\pi^3 + 26\pi^2 + 8\pi + 1$

$= (5\pi^2 + 4\pi + 1)^2$. (Yes!)

Let's also check by plugging values for a and b into the last result of Part II:

$(1 + 2\pi - \pi^2)/(\pi^2 + 2\pi + 1) = [2 - (c-1)^2]/(c+1)^2$.

Cross-multiplying, we get:

$c^2 + 2\pi c^2 - \pi^2 c^2 + 2c + 4\pi c - 2c\pi^2 + 1 + 2\pi - \pi^2$

$= \pi^2 + 2\pi + 1 + 2c\pi^2 + 4\pi c + 2c - \pi^2 c^2 - 2\pi c^2 - c^2$.

Simplifying: $2c^2 - 4c\pi^2 - 2\pi^2 + 4\pi c = 0$, $(4\pi + 2)c^2 - 4\pi^2 c - 2\pi^2 = 0$

Factoring: $[c - \pi][(4\pi + 2)c + 2\pi] = 0$, $c = \pi, -2\pi/c(4\pi + 2)$ (Yes!)

The resulting values for x, $x + a$, and $x + b$, when $c = \pi$ – which we have just submitted to a most tedious check for accuracy – are clearly *not* integral values.

Second, let's use $2^{1/2}$ (the square root of 2). (I'm terribly sorry for the awkward appearance of the fractional exponents I have to use, instead of radicals.) Suppose $x = 2^{1/2}(a + b)$, $c = 2^{1/2}$. Plugging into $x^2 + (x + a)^2 = (x + b)^2$, we get:

$$2^{1/2}(a+b) + [2^{1/2}(a+b) + a]^2 = [2^{1/2}(a+b) + b]^2$$

$$2a^2 + 4ab + 2b^2 + 2(a+b)^2 + 2a2^{1/2}(a+b) + a^2 = 2(a+b)^2 + 2b2^{1/2}(a+b) + b^2.$$

$$3a^2 + 4ab + 2(2^{1/2})(a+b)(a-b) + b^2 = 0$$

$$3a^2 + 2a^2 2^{1/2} + 4ab + b^2 - 2b^2 2^{1/2} = 0$$

$$a^2(3 + 2^{1/2}\,2) + 4ab + b^2(1 - 2^{1/2}\,2) = 0$$

$$[a(3 + 2^{1/2}\,2) + b(1 - 2^{1/2}\,2)](a+b) = 0$$

$$[a(3 + 2^{1/2}\,2) - b(2^{1/2}\,2 - 1)](a+b) = 0$$

$$a = (2^{1/2}\,2 - 1)b/(3 + 2^{1/2}\,2), -b.$$

Now, when $b = 3 + 2^{1/2}\,2$, then $a = 2^{1/2}\,2$, so:

$$x = 2^{1/2}(2^{1/2}\,2 - 1 + 3 + 2^{1/2}\,2) = 2^{1/2}(2^{1/2}\,4 + 2) = 8 + 2^{1/2}\,2$$

$$x + a = 8 + 2^{1/2}\,2 + 2^{1/2}\,2 + 1 = 2^{1/2}\,4 + 7$$

$x + b = 8 + 2^{1/2}\,2 + 3 + 2^{1/2}\,2 = 2^{1/2}\,4 + 11$, none of which is an integer-length.

Checking: $(8 + 2^{1/2}\,2)^2 + (2^{1/2}\,4 + 7)^2 = (2^{1/2}\,4 + 11)^2$

$$64 + 2^{1/2}\,32 + 8 + 32 + 2^{1/2}\,56 + 49 = 32 + 2^{1/2}\,88 + 121$$

$$153 + 2^{1/2}\,88 = 153 + 2^{1/2}\,88 \text{ (Yes!)}$$

Further checking: $2^{1/2}\,2 - 1 = [2 - (c-1)^2]/(c+1)^2 \,(3 + 2^{1/2}\,2)$

$$(c^2 + 2c + 1)(2^{1/2}\,2 - 1) = (-c^2 + 2c + 1)(3 + 2^{1/2}\,2)$$

$$2^{1/2}\,2c^2 + 2^{1/2}\,4c + 2^{1/2}\,2 - c^2 - 2c - 1 = -3c^2 + 6c + 3 + 2^{1/2}\,2 + 2^{1/2}\,4c = 2^{1/2}\,2c^2$$

$$2^{1/2}\,4c^2 + 2c^2 - 8c - 4 = 0$$

$(c - 2^{1/2})(2^{1/2} 4c + 2c + 2^{1/2} 2) = 0$

$c = 2^{1/2}$ (Yes!)

Having shown that the relationship does not hold for two particular irrational values of c, we might suspect that it fails to hold for *any* irrational value of c. But how can we verify this suspicion?

Very simply: if we assume that x is integral in length – and, therefore, rational – and that c is irrational, it follows that $x/c = (a + b)$ is irrational. (A rational number divided by an irrational number is irrational.)

From this, it follows that a or b (or both) is irrational. (An irrational number is the sum of two irrational numbers, or of one rational number and one irrational number.)

From this, it follows that $x + a$ or $x + b$ (or both) is irrational. (The sum of a rational number and an irrational number is irrational.)

This completes our proof that for any right triangle $(x, x + a,$ and $x + b)$ and for only those rational numbers c, where $c < -1$, $0 < c < (2^{1/2} + 1)$, and $(2^{1/2} + 1) < c$, the equation $x = c(a + b)$ will generate integral values for x, $x + a$, and $x + b$.

Part IV: Limits on the ratios of the sides of a right triangle

Finally, let's examine two more values for c. This will help us see yet another pattern of relationships between the sides of right triangles with integral-length sides.

Suppose $x = 100(a + b)$. Then:

$10{,}000a^2 + 20{,}000ab + 10{,}000b^2 + 200a^2 + 200ab - 200ab - 200b^2 + a^2 - b^2 = 0$

$10{,}201a^2 + 20{,}000ab + 9799b^2 = 0$

$a = -9799ab/10{,}201.$

If $a = -9799$, $b = 10{,}201$, then:

$x = 100[10{,}201 + (-9799)] = 100(402)$

$x + a = 30,401$, $x + b = 50,401$

$40,200^2 + 30,401^2 = 50,401^2$? Yes!

Or, we could have skipped the extra algebra by using thusly the formula we derived earlier:

$a = [2 - (100 - 1)2]b/(100 + 1)2 = (2 - 9801)b/10,201 = -9799b/10,201$ (etc.).

Suppose $x = 1000(a + b)$. Then:

$a = [2 - (1000 - 1)^2]b/(1000 + 1)^2 = (2 - 998,001)b/1,002,001 = -997,999b/1,002,001$.

If $a = -997,999$, $b = 1,002,001$, then:

$x = 4,002,000$.

$x + a = 3,004,001$, $x + b = 5,006,001$.

$4,002,000^2 + 3,004,001^2 + 5,006,001^2$? Yes!

Next, let's array some of the values for x, $x + a$, and $x + b$, plus some ratios:

c	1	2	3	4	10	100	1000
x	3	20	21	72	420	40,200	4,002,000
$x + a$	4	21	20	65	349	30,401	3,004,001
$x + b$	5	29	29	97	541	50,401	5,004,001
$\dfrac{x}{(x+a)}$	.75	.95	1.05	1.11	1.2	1.322	1.332
$\dfrac{x}{(x+b)}$	.6	.6896	.724	.742	.776	.798	.7994
$\dfrac{(x+a)}{(x+b)}$	.8	.724	.69	.711	.645	.603	.6003

As we know, the smallest Pythagorean triangle with integral-length sides is the 3-4-5 triangle, with the ratios $x/(x + a) = .75$, $x/(x + b) = 3/5 = .60$, and $(x + a)/(x + b) = 4/5 = .80$. These values are, respectively, minima for $x/(x + a)$ and $x/(x + b)$ and a maximum for $(x + a)/(x + b)$.

Appendix: Studies on the Pythagorean Theorem

As c increases and $x + a$ quickly becomes smaller than x, the ratios move toward what appear to be different limiting values.

Specifically, as c takes on larger and larger values (as do the lengths of the sides), the ratio of the first side to the hypotenuse goes from 3/5 to what appears to be a limit of 4/5. The ratio of the second side to the hypotenuse appears to go from 4/5 to 3/5. And the ratio of the first side to the second appears to go from 3/4 to 4/3.

But are these truly limits? Is $\lim_{c \to \infty} x/(x + b) = .80$? Is $\lim_{c \to \infty} (x + a)/(x + b) = .60$? Is $\lim_{c \to \infty} x/(x + a) = 1.3333...$? In other words:

1) Is it true that neither of the sides adjacent to the right angle can ever be less than 3/5 the size of the hypotenuse, nor more than 4/5 the size of the hypotenuse?

2) Is it true that one of the sides can never be less than 3/4 the size of the other, nor more than 4/3 the size of the other?

For those conclusions are what this limited array of values would suggest.

Secondly, what about if $c \to 0$? An array of values suggests that $x/(x + a)$ and $x/(x + b)$ also both approach 0, while $(x + a)/(x + b)$ approaches 1. In other words, as $c \to 0$, the right triangles generated appear to approach closer and closer to a state of "collapse" – i.e., a straight line, with no first side and with the second side and the hypotenuse coinciding. But are these truly limits?

Thirdly, suppose that x and $x + a$ are equal. Can the three sides of such a right triangle ever all be integral in length? Consider the following array of values:

c	1	2	2.1	2.414	2.415	2.9	3
x	3	20	2184	7,034,396	1,126,356	3944	21
$x + a$	4	21	2263	7,034,547	1,124,131	3783	20
$x + b$	5	29	3145	9,948,245	1,592,845	5465	29
a	1	1	79	151	−89	−161	−1
b	2	9	961	2,913,849	466,489	1521	9

	.75	.952	.965	.99998	1.0019793	1.041	1.05
$\frac{x}{(x+a)}$							
$\frac{a}{b}$	.5	.11	.08	.000052	−.0001907	−0.10	−0.11
$\frac{(x+b)}{(x+a)}$	1.25	1.38	1.39	1.414198	1.416956	1.44	1.45
$\frac{(x+b)}{x}$	1.67	1.45	1.44	1.414229	1.414158	1.39	1.38

This array suggests that $\lim_{c \to (1+2^{1/2})} (x + a) = x$, $\lim_{c \to (1+2^{1/2})} x/(x + a) = 1$, and $\lim_{c \to (1+2^{1/2})} x/(x + b) = 2^{1/2}$. In other words, for values of c closer and closer to $(1 + 2^{1/2})$, the values of x, $x + a$, and $x + b$ appear to become astronomically large, and the ratios of the hypotenuse, $x + b$, to the other two sides, x and $x + a$, appear to swiftly close in on $2^{1/2}$.

This suggests that any right triangle infinitesimally different from isosceles will have integral sides approaching infinity in length – and that there will, in fact, be *no actual* isosceles right triangles with all sides of integral length. But is this true?

Let's explore these three limiting cases and see if these inductive conclusions can be proved deductively.

First, consider the case when $x + a < c$, and $c \to \infty$.

$x/(x + b)$

$= c(a + b)/[c(a + b) + b]$

$= c\,([2 - (c - 1)^2]b/(c + 1)^2 + b)/$
$\quad [c\,([2 - (c - 1)^2]b/(c + 1)^2] + b) + b]$

$= bc([2 - (c - 1)^2]/(c + 1)^2 + 1)/$
$\quad b\,[c([2 - (c - 1)^2]/[(c + 1)^2 + 1) + 1]$

$= ([2 - (c - 1)^2]/(c + 1)^2 + 1)/([2 - (c - 1)^2]/(c + 1)^2 + 1 + 1/c)$

$= [2 - (c - 1)^2 + (c + 1)^2]/(c + 1)^2 \,/\, [2c - c(c - 1)^2 + c\,(c + 1)^2 + (c + 1)^2]/c(c + 1)^2$

$= (2c - c^3 + 2c^2 - c + c^3 + 2c^2 + c)/$

Appendix: Studies on the Pythagorean Theorem

$$(2c - c^3 + 2c^2 - c + c^3 + 2c^2 + c^2 + 2c + 1)$$
$$= (4c^2 + 2c)/(5c^2 + 3c + 1) = (4 + 2/c)/(5 + 3/c + 1/c^2)$$

So, $\lim_{c \to \infty} x/(x + b) = (4 + 0)/(5 + 0 + 0) = 4/5$.

$$x/(x + a) = c(a + b)/[(a + b) + a]$$
$$= c(a + [c + 1]^2 a/[2 - (c - 1)^2])/$$
$$\quad [c(a + [c + 1]^2/[2 - (c - 1)^2]) + a]$$
$$= ac(1 + [c + 1]^2/[2 - (c - 1)^2])/$$
$$\quad a[c(1 + [c + 1]^2/[2 - (c - 1)^2]) + 1]$$
$$= [1 + [c + 1]^2/[2 - (c - 1)^2]]/$$
$$\quad ([1 + [c + 1]^2/[2 - (c - 1)^2] + 1/c)$$
$$= [2 - (c - 1)^2 + (c + 1)^2]/[2 - (c - 1)^2]/$$
$$\quad (c[2 - (c - 1)^2] + c(c + 1)^2 + [2 - (c - 1)^2])/c[2 - (c - 1)^2]$$
$$= [2c - c(c - 1)^2 + c(c + 1)^2] /$$
$$\quad [2c - c(c - 1)^2 + c(c + 1)^2 + 2 - (c - 1)^2$$
$$= (2c - c^3 + 2c^2 - c + c^3 + 2c^2 + c)/$$
$$\quad (2c - c^3 + 2c^2 - c + c^3 + 2c^2 + c + 2 - c^2 + 2c + 1)$$
$$= (4c^2 + 2c)/(3c^2 + 4c + 1) = (4 + 2/c)/(3 + 4/c + 1/c^2)$$

So, $\lim_{c \to \infty} x/(x + a) = (4 + 0)/(3 + 0 + 0) = 4/3$.

$$(x + a)/(x + b) = [c(a + b) + a]/[c(a + b) + b]$$
$$= [bc + a(c + 1)]/[b(c + 1) + ac]$$
$$= ([(c + 1)^2 ac]/[2 - (c - 1)^2] + a[c + 1]) /$$
$$\quad [a(c + 1)(c + 1)^2]/[(2 - [c - 1]^2) + ac]$$
$$= ([c(c + 1)^2] + (c + 1)[2 - (c - 1)^2]) /$$
$$\quad ([c + 1]^3 + c[2 - (c - 1)^2])$$
$$= (c^3 + 2c^2 + c + 2c + 2 - c^3 + 2c^2 - c - c^2 + 2c - 1) /$$
$$\quad (c^3 + 3c^2 + 3c + 1 + 2c - c^3 + 2c^2 - c)$$
$$= (3c^2 + 4c + 1)/(5c^2 + 4c + 1) = (3 + 4/c + 1/c^2)/(5 + 4/c + 1/c^2)$$

So, $\lim_{c \to \infty} (x + a)/(x + b) = (3 + 0 + 0)/(5 + 0 + 0) = 3/5$.

Second, consider the case when $x < x + a$, and $c \to \infty$.

$\lim_{c \to 0} (x + a)/(x + b) = \lim_{c \to 0} [c(a + b) + a]/[c(a + b) + b]$

$= \lim_{c \to 0} (0 + a)/(0 + b) = \lim_{c \to 0} a/b$

$= \lim_{c \to 0} [2 - (c - 1)^2]b/(c + 1)^2 b = [2 - (0 - 1)^2]/(0 + 1)^2 = 1$.

$\lim_{c \to 0} x/(x + a) = \lim_{c \to 0} c(a + b)/[c(a + b) + a] = 0/(0 + a) = 0$.

$\lim_{c \to 0} x/(x + b) = \lim_{c \to 0} c(a + b)/[c(a + b) + b] = 0/(0 + b) = 0$.

Third, consider the case when $a = 0$, $x = x + a$, and the triangle is isosceles.

$x - c(a + b) = c(0 + b) = cb$, $c = x/b$.

$(x + b)^2 = x^2 + (x + a)^2 = x^2 + x^2 = 2x^2$.

$x + b = 2^{1/2} x$.

$b = x(2^{1/2} - 1)$.

$c = x/x(2^{1/2} - 1) = 1/(2^{1/2} - 1) = (2^{1/2} + 1)/(2^{1/2} - 1)(2^{1/2} + 1)$
$= (2^{1/2} + 1)/(2 - 1) = 2^{1/2} + 1$.

$(x + a)/x = 1$ (isosceles triangle).

$(x + b)/x = [c(0 + b) + b]/c(0 + b) = (bc + b)/bc$
$= 1 + 1/c = 1 + 1/(2^{1/2} + 1)$
$= (2^{1/2} + 1 + 1)/(2^{1/2} + 1) = (2^{1/2} + 2)/(2^{1/2} + 1)$
$= (2^{1/2} + 2)(2^{1/2} - 1)/(2^{1/2} + 1)(2^{1/2} - 1)$
$= (2 + 2^{1/2} - 2)/(2 - 1) = 2^{1/2}$.

$(x + b)/(x + a) = (x + b)x/(x + a)x$
$= [(x + b)/x][x/(x + a)] = 2^{1/2}(1) = 2^{1/2}$.

This proves that the ratio of the hypotenuse to either of the sides of an isosceles right triangle will always be $2^{1/2}$, and that there can be no such triangle with all three sides of integral length. If the length of the hypotenuse $(x + b)$ is some integer, m, then the lengths of the other two sides will be $[(1/2)m^2]^{1/2}$, or $m/2^{1/2}$, or $2^{1/2}m/2$.

If, on the other hand, the length of the smaller two sides is some integer, n, then the length of the hypotenuse is $(2n^2)^{1/2}$, or $n(2)^{1/2}$. In the former case, the length of the hypotenuse is non-integral. In the latter, the lengths of the two smaller sides are non-integral.

We now also know precisely why the values for x and $x + a$ "flip-flop" between $c = 2$ and $c = 3$. The values for a are positive between for $0 < c < 2.41421356+$, i.e., for $0 < c < (2^{1/2} + 1)$, 0 when $c = (2^{1/2} + 1)$, and negative when $(2^{1/2} + 1) < c$.

Now we can see not only that the values for x and $x + a$ flip-flop at this point, but also that they become astronomically large just below and above the flip-flop point.

Conclusion

What can we say about Fermat's Last Theorem, based upon the findings of this paper? We know that when $n = 2$, $x^n + y^n = z^n$ has *real* solutions, because there are real solutions for $x = [2b (b - a)]^{1/2} + (b - a)$, $x = c (a + b)$, and $a = [2 - (c - 1)^2]b/(c + 1)^2$.

While many values of n have been tried for $x^n + y^n = z^n$ and found wanting, this study was not successful in showing that the Pythagorean Equation *cannot* be true for any $n > 2$. That proof was finally provided in 1995 by Andrew Wiles and Richard Taylor, putting the mystery to rest once and for all.

What we *do* know from this study that we did not know before, however, is a formula for generating the infinitude of integral solutions of the Pythagorean equation – and the limits within which those solutions exist.

These are certainly results that would not have occurred to this writer, without having first set out to work on the Fermat conjecture and followed the inductive trail through a lot of detailed calculation and thought.

Postscript: comparison of the author's method with other formulas for generating Pythagorean triples

Since writing the first three sections of this essay, I have discovered two other formulas that have been used for generating "Pythagorean triples" – i.e., integral solutions for the Pythagorean equation. One was devised by the ancient Pythagoreans (or perhaps adopted by them from the Egyptians and/or Babylonians), while the other appears to be of more recent origin:

1) Given odd integer, r, a Pythagorean triple (x, y, z) is generated by setting $x = r$, $y = (1/2) (r^2 - 1)$ and $z = (1/2) (r^2 + 1)$. (It can be seen

that these conditions satisfy the Pythagorean equation: $r^2 + [(1/2)(r^2 - 1)]^2 = [(1/2)(r^2 + 1)]^2$, so $r^2 + (1/4)r^4 - (1/2)r^2 + 1/4 = (1/4)r^4 + (1/2)r^2 + 1/4$.

2) Given two integers, m and n, where $m > n > 0$, $m + n$ is odd, and m and n have no common factor, a Pythagorean triple (x, y, z) in which x is odd can be generated by setting $x = m^2 + n^2$, $y = 2mn$, and $z = m^2 + n^2$. (It can be seen that these conditions satisfy the Pythagorean equation: $(m^2 - n^2)^2 + (2mn)^2 = (m^2 + n^2)^2$, so $m^4 - 2m^2n^2 + n^4 + 4m^2n^2 = m^4 + 2m^2n^2 + n^4$.)

Compare these equations and procedures with the ones I have defined in this paper: any rational number, c, determines a Pythagorean triple (x, $x + a$, $x + b$), in which $x = c(a + b)$ and $a/b = [2 - (c - 1)^2]/(c + 1)^2$. Solve the second equation for a/b, set the numerator equal to a and the denominator equal to b, solve the first equation for x, thereby determining the values for $x + a$ and $x + b$. The procedure is more complicated, but the initial input is simplicity itself: one rational number (greater than 0 or smaller than −1).

Comparison of the author's method with procedure (1): as a further comparison between my procedure and (1) above, consider that:

1) When $c = 1$, my procedure arrives at (3, 4, 5), which is what procedure (1) arrives at when $r = 3$.

2) When $c = 1/3$, my procedure arrives at (5, 12, 13), which procedure (1) arrives at when $r = 5$.

3) When $c = 1/5$, my procedure arrives at (7, 24, 25), which procedure (1) arrives at when $r = 7$.

4) When $c = 1/7$, my procedure arrives at (9, 40, 41), which procedure (1) arrives at when $r = 9$.

The relationship between the inputs of these two procedures is empirically obvious: the r inputs are the odd numbers starting with 3, while the c inputs are the fractions with a numerator of 1 and an odd-numbered denominator such that, for a given triple, the denominator of c is 2 smaller than r. In other words, for a given triple, $c = 1/(r - 2)$, where r is an odd number, $r > 1$.

This relationship can also be determined algebraically (i.e., rationally), as

Appendix: Studies on the Pythagorean Theorem

follows:

$$x = r,\ x + a = (r^2 - 1)/2,\ x + b = (r^2 + 1)/2.$$

$$r + a = r^2 - 1/2,\ \text{so}\ a = r^2/2 - r - 1/2.$$

$$r + b = r^2 + 1/2,\ \text{so}\ b = r^2/2 - r + 1/2.$$

$$x = c\,(a + b)$$

$$r = c\,(r^2/2 - 1/2 - r + r^2/2 + 1/2 - r) = c\,(r^2 - 2r)$$

$$c = r/(r^2 - 2r) = 1/(r - 2)$$

As a check, substitute into $a/b = (1 + 2c - c^2)/(1 + 2c + c^2)$:

$$(r^2/2 - r - 1/2)/(r^2/2 - r + 1/2)$$
$$= [1 + 2/(r-2) - 1/(r-2)^2]\,/\,[1/(r-2)^2 + 2/(r-2) + 1]$$
$$= [(r-2)^2 + 2(r-2) - 1]\,/\,[1 + 2(r-2) + (r-2)^2]$$
$$= (r^2 - 4r + 4 + 2r - 4 - 1)\,/\,(1 + 2r - 4 + r^2 - 4r + 4)$$
$$= (r^2 - 2r - 1)/(r^2 - 2r + 1)$$
$$= [r^2/2 - r - 1/2]\,/\,[r^2 - r + 1/2]\quad \text{(Check!)}$$

It is also clear how limited procedure (1) is, compared with my procedure, in generating triples. Mine generates all the triples that procedure (1) generates with the odd integers – plus the infinity of triples produced by c being allowed to take on the values of all the rational numbers larger than 0 and smaller than –1, none of which procedure (1) is able to produce.

Comparison of the author's method with procedure (2): It is probably somewhat easier to derive the relationship between the inputs of procedure (2) and my procedure by the rational (deductive, algebraic) approach:

$$x = m^2 - n^2,\ x + a = 2mn,\ x + b = m^2 + n^2$$

$$m^2 - n^2 + a = 2mn,\ \text{so}\ a = 2mn - m^2 + n^2$$

$$m^2 - n^2 + b = m^2 + n^2,\ \text{so}\ b = 2n^2$$

$$x = c\,(a + b)$$

$$m^2 - n^2 = c\,(2mn - m^2 + n^2 + 2n^2)$$
$$= c\,(2mn - m^2 + 3n^2)$$

$$= c(3n^2 + 2mn - m^2)$$
$$c = (m^2 - n^2)/(3n^2 + 2mn - m^2)$$
$$= (m + n)(m - n)/(3n - m)(n + m)$$
$$= (m - n)/(3n - m)$$

As a check, substitute into $a/b = (1 + 2c - c^2)/(1 + 2c + c^2)$:

$$(2mn - m^2 + n^2)/2n^2$$
$$= (1 + 2[(m - n)/(3n - m)] - [(m - n)/(3n - m)]^2) /$$
$$\quad ([(m - n)/(3n - m)]^2 + 2[(m - n)/(3n - m)] + 1)$$
$$= [(3n - m)^2 + 2(m - n)(3n - m) - (m - n)^2] /$$
$$\quad [(m - n)^2 + 2(m - n)(3n - m) + (3n - m)^2]$$
$$= (9n^2 - 6mn + m^2 + 6mn - 2m^2 - 6n^2 + 2mn + 9n^2 - 6mn - n^2) /$$
$$\quad m^2 - 2mn + n^2 + 6mn - 2m^2 - 6n^2 + 2mn + 9n^2 - 6mn + m^2$$
$$= (2n^2 - 2m^2 + 4mn)/4n^2$$
$$= (n^2 + 2mn - m^2)/2n^2. \text{ (Check!)}$$

Procedure (2) is a much more powerful triple generator than procedure (1), as can be seen from the following chart, comparing its output with that from my procedure. It is capable of generating not only all the triples produced by procedure (1), but also all the other triples in which x is an odd number:

c	$m^2 - n^2$	$2mn$	$m^2 + n^2$	(m, n)
1	3	4	5	(2,1)
1/3	5	12	13	(3,2)
1/5	7	24	25	(4,3)
1/7	9	40	41	(5,4)
1/9	11	60	61	(6,5)
3/–1	15	8	17	(4,1)
3	21	20	29	(5,2)

[note: 3/3 aka (6, 3) omitted: yields (27, 36, 45) = (3, 4, 5)]

3/5	33	56	65	(7,4)

Appendix: Studies on the Pythagorean Theorem

3/7	39	80	89	(8,5)	

[note: 3/9 aka (9, 6) omitted: yields (45, 108, 117) = (5, 12, 13)]

3/11	51	140	149	(10,7)	
5/–3	35	12	37	(6,1)	
–5	45	28	53	(7,2)	
5/3	55	48	73	(8,3)	

[note: 5/5 aka (10, 5) omitted: yields (75, 100, 125) = (3, 4, 5)]

7/–5	63	16	65	(8,1)	
7/–3	77	36	85	(9,2)	
–7	91	60	109	(10,3)	

The values for c and (m, n) containing common factors were included above, in order to show in an unbroken manner the progression of values in the numerator and denominator (henceforth, g and h) of c, and in the pairs (m, n). From this, it can be seen that: $c = (g/h) - (m - n)/[m - (m - n) + n - (m - n)] = (m - n)/(3n - m)$.

As does my procedure, procedure (2) creates "flip-flop" triples – i.e., triples in which the second term is smaller than the first. Unlike my procedure, however, procedure (2) gives no clear indication as to where or why the flip-flop occurs (namely, at $c = 2^{1/2} + 1$, and $-1 < c < 0$.

Therefore, if the flip-flop triples are generated when $c > 2^{1/2} + 1$, and when $c < -1$, it follows that they must occur when $(m - n)/(3n - m) > 2^{1/2} + 1$, and when $(m - n)/(3n - m) < -1$.

It is hard to see how this conclusion could be reached merely by considering procedure (2) in itself. Having developed my own procedure and comparing it to procedure (2) is likely the only way the flip-flop points could have been identified for procedure (2).

Granted, procedure (2) does not put a priority on distinguishing between the domain of the triples in which $x < x + a < x + b$ (namely, $0 < c < 2^{1/2} + 1$) – versus that in which $x + a < x < x + b$, (namely, $2^{1/2} + 1 < c$, and $c < -1$.) Instead, it puts a priority on generating triples whose first term is odd.

It can, however, easily be redefined so as to generate triples with an even first term by setting $x = 2mn$, $x + a = m^2 - n^2$, and $x + b = m^2 + n^2$.

My procedure, of course, does this equally well. When *either* the numerator or denominator (if any) of c is even, the first term of the triple it generates is even, also. If the numerator and denominator of c are *both* odd, the first term of the triple is odd. Here are a few examples to illustrate:

c	x	$x + a$	$x + b$
1	3	4	5
→inf	4	3	5
1/3	5	12	13
2/–1	12	5	13
1/2	8	15	16
3/–1	15	8	17
1/5	7	24	25
–3/2	24	7	25
2	20	21	29
3	21	20	29
1/4	12	35	37
5/–3	35	12	37
1/7	9	40	41
–4/3	40	9	41
1/9	11	60	61
–5/4	60	11	61
1/6	16	63	65
7/–5	63	16	65
5	55	48	73
3/2	48	55	73
4	72	65	97

5/3	65	72	97	
1/8	20	99	101	
9/–7	99	20	101	
7	105	88	137	
4/3	88	105	137	
6	156	133	205	
7/5	133	156	205	
9	171	140	221	
5/4	140	171	221	
8	272	225	353	
9/7	225	272	353	

The preceding table also provides some clues about the relationship between the values for c that generate triples with the first term smaller than the second and the values for c (henceforth c_f) that generate the respective flip-flop triples. Following is a re-ordering of the above data, with some additional data that will help in seeing the relationship we are looking for:

c	c_f
1/3	2/–1
1/2	3/–1
1/5	3/–2
1/4	5/–3
1/7	4/–3
1/6	7/–5
1/9	5/–4
1/8	9/–7
5/3	4
3/2	5

7/5	6
4/3	7
21/10	31/11
39/19	29/10

As a result of a trial-and-error study of this array, the most likely candidate is for c_f to be equal to the sum of g (the numerator of c) and h (the denominator of c) divided by the difference of g and h – or $(g_c + h_c)/(g_c - h_c)$. Although I have not validated it, this relationship fits the above wide variety of examples of flip-flops.

We can now identify the relationship between triples that are in a flip-flop relationship to those related to the pair (m, n) by the equation $c = (m - n)/(3n - m)$:

$$c_f = (g_c + h_c)/(g_c - h_c) = [(m - n) + (3n - m)]/[(m - n) - (3n - m)] = 2n/(2m - 4n) = n/(m - 2n)$$

As a result, we now have equations expressing the relationship between the two procedures, so that c can be derived from (m, n) in either case:

(1) <u>When x is odd</u>: $c = g/h = (m - n)/(3n - m)$

(2) <u>When x is even</u>: $c = (g + h)/(g - h) = n/(m - 2n)$

Furthermore, we can solve these equations and derive two more equations by which we can derive (m, n) from c in either case:

(1) <u>When x is odd</u>: $g/h = (m - n)/(3n - m)$, so $mh - nh = 3ng - mg$, $m(h + g) = n(3g + h)$, $m/n = (3g + h)/(g + h)$

(2) <u>When x is even</u>: $m/n = [3(g + h) + (g - h)]/[(g + h) + (g - h)] = (4g + 2h)/2g = (2g + h)/g$

Here is an example to illustrate how these four transformation equations work:

> If $c = 21/10$, $x = 2184$ (which is even), so $m/n = [2(21) + 10]/21$, $21m = 42n + 10n$, $m/n = 52/21$, so $(m, n) = (52, 21)$. Because x is even, $c = 21/[52 - 2(21)] = 21/(52 - 42) = 21/10$, and $c_f = (52 - 21)/[3(21) - 52] = 31/(63 - 52) = 31/11$.

Appendix: Studies on the Pythagorean Theorem

If $c = 7/-1$, $x = 91$ (which is odd), so $m/n = [3(7) + (-1)]/[7 + (-1)]$
$= (21 - 1)/(7 - 1) = 20/6 = 10/3$, so $(m, n) = (10, 3)$. Because x is
odd, $c = (10 - 3)/[3(3) - 1] = 7/(9 - 10) = 7/-1$, and $c_f = 3/(10 - 2(3))] = 3/(10 - 6) = 3/4$.

One last tidbit: a serendipitous result of this last example is the insight that the flip-flop value for the *negation* of c is the *inverse* of the flip-flop value for c. (Compare 7/-1 and 3/4 with 7 and 4/3.) The proof is simple:

$c_f = (g + h)/(g - h)$, so $1/c_f = (g - h)/(g + h)$

$-c = -g/h$, so $-c_f = (-g + h)/(-g - h) = (g - h)/(g + h)$

Therefore, $1/c_f = -c_f$, and thus $c_f = 1/-c_f$.

Following is a table of values for c, $c_f = 1/-c_r$, $-c$, and $-c_f = 1/c_f$, to illustrate these final equations:

c	$c_f = 1/-c_f$	$-c$	$-c_f = 1/c_f$
2	3	2/–1	1/3
3	2	3/–1	1/2
4	5/3	4/–1	3/5
5	3/2	5/–1	2/3
6	7/5	6/–1	5/7
7	4/3	7/–1	3/4
8	9/7	8/–1	7/9
9	5/4	9/–1	4/5
1/1	1/–0 (infinite)	1/–1*	0/–1*
1/2	3/–1	1/–2*	1/–3*
1/3	2/–1	1/–3*	1/–2*
1/4	5/–3	1/–4*	3/–5*
1/5	3/–2	1/–5*	2/–3*
1/6	7/–5	1/–6*	5/–7*
1/7	4/–3	1/–7*	3/–4*
1/8	9/–7	1/–8*	7/–9*
1/9	5/–4	1/–9*	4/–5*

Note that the asterisked (*) numbers are values for which the triples generated have one or more negative values. These correspond to right triangles with one or more sides having a "negative length."

These are entities which, though they still satisfy the Pythagorean Equation, are purely imaginary. In this respect, the values where $0 < c < 1$ and $c_f < -1$ are not as "fertile" as the values where $1 \leq c < 2^{1/2} + 1$, and $2^{1/2} + 1 < c_f$. In the latter cases, the negation of c and the inversion of c_f generate real right triangles, while in the former cases, they do not.

Finally, here is the text of the cover letter I sent out to several mathematicians in the Southern California area. I got no significant feedback on the essay itself, but I did get some suggestions about possible journal publication, which I decided not to pursue at the time.

> October 22, 1992
> Dear Reader:
>
> The attached is the work of an amateur mathematician. I turned away from this field many years ago, in order to become a professional musician. Recently, however, I have been inspired by Fermat's Last Theorem and by the life and work of Ramanujan, the great Indian mathematician of the early 1900s. One of the results is now in your hands.
>
> The approach I have used throughout is intuitive induction followed by logical deduction. I wanted both to illustrate and document my approach, and then to validate it. For your convenience, I have also provided a three-page condensed version of my principal results and their proofs.
>
> I do not know whether what I have presented is original or is merely an independent "re-inventing of the wheel." Naturally, if I have been the first to arrive at these conclusions about the Pythagorean Theorem, I would like to know – and to receive recognition for having done so.
>
> Your judgment of this work, and your suggestions for what to do with it next, would be greatly appreciated. Thank you.
>
> Sincerely,
> (Signed)
> Roger Bissell
> (Orange, California address and phone included)

www.ingramcontent.com/pod-product-compliance
Lightning Source LLC
Chambersburg PA
CBHW030939180526
45163CB00002B/636